Day Skipper Sailing

Karle Stephenson

The Crowood Press

First published in 1998 by
The Crowood Press Ltd
Ramsbury, Marlborough
Wiltshire SN8 2HR

British Library Cataloguing in Publication Data

A catalogue record for this book is available from the British Library.

ISBN 1 86126 034 2

Line drawings by Andrew Green

The Macmillan Silk Cut Nautical Almanac 1994, reprinted by kind permission of
Pan Reference Books

Chartwork reproduced by permission of the Controller of Her Majesty's
Stationery Office and the UK Hydrographic Office

Typeset by Annette Findlay
Printed by J.W. Arrowsmith, Bristol

CONTENTS

1

TERMINOLOGY

To the novice, the sailor's world seems full of strange new words and expressions. Some of these are polite and repeatable on shore, others are used in moments of 'serious sea-going stress', and are uttered to reassure the crew that, whatever the problem may be, it is in fact quite trivial. It also indicates that the skipper is not the least concerned about the two feet of water in the saloon, or the fact that the spinnaker is tightly wrapped around the forestay and looks like staying that way.

Every single part of a yacht and all the equipment associated with sailing are there for a specific purpose, and all have identifying names which are important and should be learned and used. For a skipper to call in the general direction of a group of crew members to 'Pull on that rope' when they are surrounded by ropes, is quite pointless, and the call will only have to be repeated several times, each time probably in a louder voice by an increasingly apoplectic skipper to an increasingly perplexed crew. Such a scene may be amusing to a third party, but it is not to the crew, and it may be downright dangerous at sea when a particular rope needs to be hauled tight in a hurry.

In this first chapter, nautical terms and expressions in everyday use at sea will be explained, and the beginner will do as well to learn them, because the prudent skipper will always use them.

Deck Area

When looking towards the front or bow of the yacht, the right-hand side of the boat

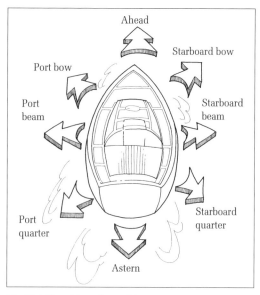

Fig 1(a) Topsides of a yacht.

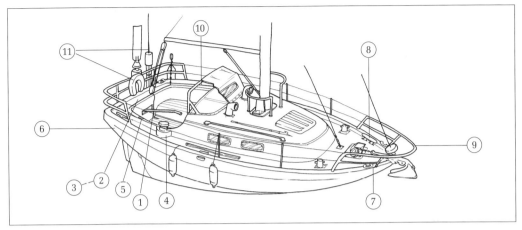

Fig 1(b) Parts of a yacht.

is the starboard side, and the left hand is the port side, or as it used to be called, the larboard side. You will read later that at night, these sides are identified to any approaching vessel by special red (port) and green (starboard) lights.

The middle section is known as 'amidships' or the beam of the boat. Forward of

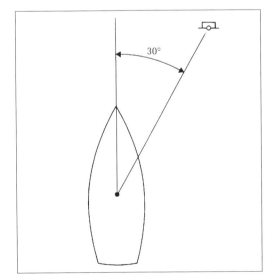

Fig 2 A Buoy shown 'bearing 030° off the starboard bow'. Note:- not to scale.

this is the foredeck, culminating in the bow of the vessel. When looking from amidships over the bow of the boat you are said to be looking 'ahead' at the ship's 'heading'. The expressions 'starboard' or 'port bow' are used to allow the position of objects such as buoys to be specified relative to the ship's heading, for example: 'A buoy 30 degrees off the port or starboard bow' (as shown in Fig. 2).

To the rear of amidships, or 'abaft' (behind) the beam, is the cockpit area in which the tiller and cockpit locker(s) (Fig. 1b (1) and (2)) are found. You don't have cupboards on yachts, you have lockers; for example, a cockpit locker is situated under the cockpit seat (Fig. 1b (3)). A locker is a deep, cavernous space used for stowing all manner of equipment, such as ropes for tying the vessel to a pontoon or jetty (when used for this purpose a rope is called a 'warp') and also ropes used to control the sails (called 'sheets').

Associated with the cockpit area are winches and cleats (Fig. 1b (4) and (5)): the winches help to haul a rope really tight, and the cleats to secure it under tension. A winch consists of a metal

Cockpit winches and jamming cleats.

drum designed to rotate freely in only one direction, usually clockwise when viewed from above. In use, the end of a rope – known as the bitter end – is passed three times around the drum, in the direction of free movement. The rope is hauled in tight, and once tensioned in this way, the friction between the three turns and the winch drum is such that only a light hold on the bitter end will suffice to retain the rope tension, at which point it may be cleated off. Care must be taken to avoid a 'riding turn' when hauling a rope in this way. Where a lot of rope is to be hauled in, it is better to start with only one turn round the winch, moving to three turns as the rope length still to be hauled is reduced. A riding turn occurs when that part already under strain rides up over any or all of the turns around the winch

drum, and it should be stressed that a riding turn can be extremely difficult to release.

Behind the cockpit is the stern or transom, (Fig. 1b (6)). At night a special white light identifies the stern of your yacht to any vessel which may be overtaking. When looking over the stern of your boat you are said to be looking 'astern'. The sides of the boat to the port and starboard side of the cockpit are referred to as the port and starboard quarters.

Deck Fittings

The Tiller

The tiller, (Fig. 1b (1)) is used to steer the vessel when under sail or under engine. However, you need to be aware that the boat must be moving through the water before the tiller has any turning affect on the yacht. The tiller is pivoted so that, for example, as you pull the tiller to port, the rudder blade is turned to starboard, this starboard movement of the blade deflecting the water streaming past it, causing the boat to turn to starboard. The faster the stream, the bigger is the deflecting power of the rudder.

Please notice that the movement of the stream of water over the rudder is relative; thus if the boat is under power and moving through water which itself is stationary over the ground, the tiller will bite and you will have steerage way. You will also have steerage way if you are heading into a two-knot stream with the boat's speed adjusted to make it stationary over the ground; this is because the water is streaming past the tiller blade at a relative speed of two knots. This action is called 'stemming the tide', and you will see later how useful this can be; for example, it would give the skipper thinking time when the yacht is being manoeuvred in a busy harbour.

When using the tiller to alter the boat's heading, you will learn to return the tiller to its middle position as the new course comes on, otherwise the boat continues to turn and will circle indefinitely while the tiller is off-centre.

The Wheel

On larger yachts the tiller is often replaced by a wheel which acts exactly as does the steering wheel of a car; therefore turning the wheel to starboard, right hand down a bit, turns the boat to starboard. However, whether it is under tiller or wheel, a yacht when turning, unlike a car, actually pivots about some central point –

that is, if the helm turns the boat to starboard the bow pivots to starboard, about the central point of the boat, and the stern pivots to port. It follows that whichever way you want the boat to turn, you must allow sea-room for the stern to pivot. On some occasions, usually when leaving or arriving at a harbour, it will be necessary to manoeuvre in quite tight places under engine, and whenever this need arises you should be aware that most boats pivot in a tighter circle in one direction than the other. This effect is in the main due to the propeller walking 'prop-walk' through the water.

Prop-walk

Propellers are manufactured as either right- or left-handed: that is, when viewed from the stern of a vessel moving forwards a right-hand propeller runs clockwise, and in astern propulsion it would run counter-clockwise. Propellers are also much more efficient when going ahead, giving a lot more thrust when going ahead than when going astern. To take advantage of prop-walk in a tight turn, the boat should be stopped or moving very slowly.

Let us assume the necessity of a tight turn in a yacht with a right-hand prop: the boat, preferably stopped in the water, is given a hard, short burst on the engine as the wheel is turned to starboard. The boat moves forwards, but although she will pivot to starboard under the tiller's influence, prop-walk will pivot the stern to starboard tending to cancel the effect of the tiller; this is the reason for using only a short sharp burst of engine power to get the boat moving. Prop-walk only occurs when the engine is providing power to the propeller; once moving forwards the tiller will continue to pivot the boat to starboard.

To complete the turn before hitting anything in this tight situation, the engine is put into reverse with a longer sharp burst, and the wheel is turned down to port. Now prop-walk and rudder blade are pivoting the boat in the same

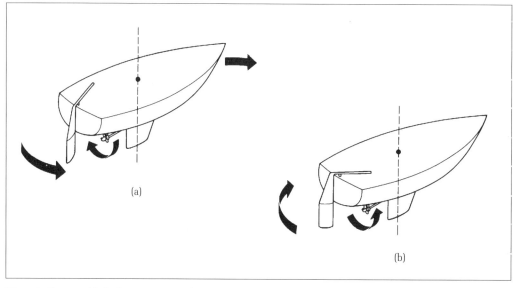

(a)

(b)

Fig 3 In diagram (a) the boat is moving forward under engine. The tiller is pivoted to port (rudder to starboard), the rudder blade distorts the flow of water such that the force of water acting on the rudder blade will cause the boat to turn to starboard.

Note:- (1) The boat actually pivots about some central point (X) ie the bow moves to starboard while the stern moves to port.

(2) The faster the flow of water over the rudder, the greater the force turning the boat, and the quicker she'll turn. It should be noted from this that as the speed of a boat is reduced, the effect of both wind and tide may become more important than the tiller in determining her heading since the rudders control will be reducing.

The diagram also shows how a 'right hand propeller', that is, a prop rotating clockwise when viewed from astern in forward propulsion, will 'walk' the stern to starboard, thereby countering the tiller's action.

Diagram (b) The boat is going astern under engine, the tiller, now pivoted to starboard, will move the stern to port and the reverse prop will walk the stern to port; prop and rudder blade are working together the stern of the boat will pivot quickly to port in a tight turn.

Note:- (1) Prop-walk is most noticeable when the boat is stationary or moving very slowly in the water.

(2) Prop walk is more powerful when going astern since the prop is less efficient in astern propulsion than it is in forward propulsion.

When attempting to perform a tight turn in a narrow channel, or when manoeuvring in a marina, the act should take advantage of prop-walk, prop-walk is more effective when going astern. Therefore arrange it so that when going astern both prop and rudder are moving the stern in the same direction. This means that for a right hand prop the initial forward turn of the bow should be to starboard. A short powerful burst of engine with the tiller to port will get the boat moving; quickly putting the engine into neutral will kill prop-walk and the boat will turn under the tiller's control. Reversing the rudder, with a longer powerful burst of the engine in reverse, will stop all forward movement: tiller and prop-walk in combination will put the stern heavily to port. Repeating these two operations will turn most yachts in their own length.

direction, ie the stern is pivoting to port. Due to the reduced efficiency of the prop when going astern, the yacht will make only a little stern-way, and most of the pivoting effort comes from the prop-walking through the water. Using this technique, most boats can be turned in a very tight circle, if not within their own length. Fig. 3 shows the pivoting effect of rudder blade and prop in a tight turn manoeuvre.

Unfortunately some boats do not respond in the above manner when going astern. It is therefore prudent, as with most operations at sea, to experiment in clear water before relying on prop assistance in a tight turn.

Fairleads and Cleats

Fairleads and cleats are usually fitted at the port and starboard bow and quarters (see Fig. 1b (7) and (8)). These fittings are used when securing the boat alongside a mooring pontoon.

In preparation for bringing a boat alongside a pontoon, the skipper will call 'Warps and fenders port (or starboard) side to', at which instruction the knowledgeable crew will proceed as follows: warps and fenders are taken from the cockpit locker, and the fenders are secured using a round turn and two half-hitches (see Rope-Work later) along the port-side handrail at a height suitable to protect the boat's side when she is moored to the pontoon. The warps are secured to the forward and aft cleats on the port side and led through the associated fairleads. One crew member to each of these warps, ensuring that the warp has been taken out through the fairlead, will bring the rope back to the shrouds.

The skipper stops the boat neatly alongside the jetty, the two crew members step – not jump – onto the jetty, and each makes a full round turn of the warp on a pontoon bollard sited opposite the bow and stern of the yacht. The boat is now stopped safely alongside, and the skipper will issue instructions to complete the mooring operation.

Mooring Alongside

When mooring a yacht to a floating pontoon, the boat is stopped alongside using two warps, as discussed above. It is then normal practice to tie bowlines (see Rope-Work, later) at the shore end of these warps, drop them over the pontoon bollards, haul any slack warp back inboard, and re-secure the warp to the fore and aft cleats. These two lines are now the fore and aft breast ropes. The breast ropes are sufficient to hold the boat alongside the pontoon, but they will not prevent her from moving forwards and backwards with tide changes, nor from surging along the jetty as other boats pass by. To prevent this fore/aft movement the skipper will rig fore and aft springs. Springs are usually two long warps, one leading from the bow through the fairlead to a shore bollard situated towards the stern of the vessel, the other leading from a cockpit winch through the fairlead to a second shore bollard sited near the bow of the vessel. These crossing springs, when tightened by means of the cockpit winch, will prevent surge and will also keep the yacht slightly off the pontoon. A fully secured yacht is shown in Fig. 4.

When mooring alongside a town quay or any other non-floating shore fixture, the change in sea level, or range of tide, will have to be considered. Short breast

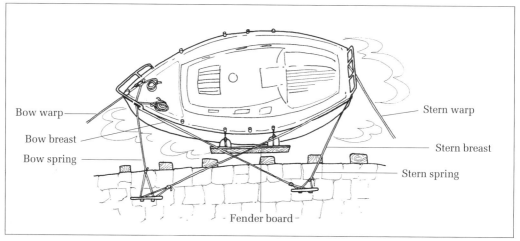

Fig 4 Fully secured alongside.

ropes and springs would need constant attention as the sea level changes with the tide; alternatively these could be replaced with longer warps run out well ahead of and behind the yacht, and the springs extended such that each line's length were some three times the expected change in the sea level.

Pulpit, Pushpit and Handrails

The stainless-steel railings at the fore and aft extremes of the boat are known as the pulpit at the bow (Fig.1b (9)) – because that's what they look like – and the push-pit at the stern (Fig. 1b (11)); pushpit is an Americanism for what was previously called the stern-rail or quarter-rail. The handrail wires are run between these two fixtures, supported at intermediary points by stainless-steel stanchions. It must be emphasized here that, although the handrails offer support to a crew member moving about the side decks under balmy sea conditions and in harbour, they should not be considered a strongpoint and may not prevent an unbalanced yachtsman going over the side in rough seas. For this reason the handrails must not be used to secure safety harness straps in heavy weather.

Jackstays

Jackstays are lengths of strong webbing or wire rope running along the side decks for the whole length of the boat. Crew members moving about the deck clip their safety harness to these lines in rough weather, at night or at any time that the skipper or they themselves deem it to be advisable to wear safety harnesses.

Anchors

The bower or main yacht anchor is normally stowed in the forepart of the boat, either secured on deck or housed with its chain in a special anchor locker ready for fast deployment.

Anchoring is an important subject and will be discussed more fully later, under that heading. A kedge or second anchor, normally stowed in the cockpit locker together with its own rope warp, is somewhat lighter than the bower, its purpose being to allow the skipper to haul his vessel from one point to another without the use of sail or engine. In use the anchor warp is secured to a cleat, and the kedge is taken away in the dingy and let go at a suitable point. The yacht is then manhandled up to the kedge. One very common use of this anchor is to 'kedge off' when a yacht is hard aground. The kedge is let go in deeper water and the boat is either literally dragged off the shoal until she refloats, or the crew wait until the rising tide refloats her, in which case the kedge simply stops her being bumped shorewards by the rising tide.

Dan-buoy and Life-buoy

These two life-saving pieces of equipment are normally mounted on purpose-built brackets attached to the pushpit (Fig. 1b (11)). The dan-buoy is a bottom-weighted telescopic pole with a small flag attached to the top, weighted thus so that it floats vertically when heaved overboard. The life-buoy, usually connected to the dan-buoy by a long buoyant line, has a waterproof light attached to it which is held upside down in the bracket, but which illuminates immediately when placed upright.

Should anyone fall overboard, these two pieces are removed from their brackets, the dan-buoy is extended and the buoyant line uncoiled, and both are thrown towards the person in the water: please note, '*towards*' rather than 'at'. The buoyant line is an extremely good idea because it lies out on the sea surface and so the person in the water need only swim towards the area between the two pieces of floating equipment so as to

Dan-buoy and life-belt mounted on pushpit.

make contact with the line before pulling in the life-buoy. The life-buoy light should be tested at regular intervals since, as a result of being exposed to the sea air and occasionally to sea water, corrosion is possible. The crew must know how to release and deploy these two pieces of kit; their use would normally form part of the skipper's pre-sail safety chat.

Masts and Rigging

The purpose of a mast and rigging is to support the sails and to transfer the driving force created by the action of the wind on the sails to the yacht's hull. The majority of yacht masts today are extruded to a designed sectional shape from seawater-resistant aluminium alloys.

The mast is supported by so-called 'standing rigging', usually made of stainless-steel rope; the components of standing rigging are as follows:

(1) Fore-and-aft stays, ie one forestay and one or two backstays;

(2) Lateral rigging, called shrouds. The shrouds are brought down from the top and upper sides of the mast to the port and starboard sides of the deck. Compression strut spreaders, in the past called crosstrees, are fitted to provide a greater and therefore more efficient angle between mast and shrouds. These components are shown in Fig. 5.

Other mast components are the boom, to which the foot of the mainsail is attached; the kicking strap or boom-vang used in sail control; and a spinnaker pole, used to control a special sail called a spinnaker. The spinnaker pole may also be used to 'pole out' the jib.

The control of the yacht's sails is carried out using a number of ropes variously described as 'running rigging'; however, we shall first describe the sails themselves, and identify their many parts.

The Sails

The most common type of modern yacht is the so-called Bermudan sloop. This type of boat has a single mast to which two triangular sails may be attached, one in front of the mast and known as the jib, foresail or headsail, and a second behind the mast and known as the mainsail.

The Jib Sails

The jib is the name given to that triangular piece of modern material which is attached to the forestay, hauled aloft by a part of the running rigging called the forehalyard, and controlled by two other ropes referred to as foresail sheets; the several parts of a foresail are shown Fig. 6a. The leech is usually fitted with a thin cord running its full length in a thin sleeve; this cord allows for very fine control of this part of the sail. A yacht may carry several jibs:

(1) A storm jib, small in size and made from heavy material.

(2) A working jib, bigger and of lighter material than the storm jib.

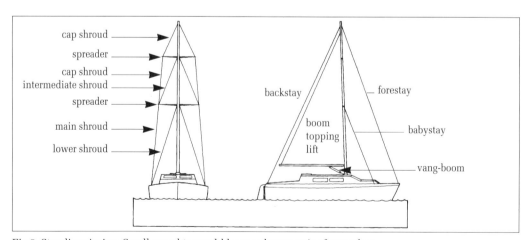

Fig 5 Standing rigging. Smaller yachts would have only one pair of spreaders.

(3) A genoa(s); these large and lighter sails, used at the lower end of wind strengths, provide large overlaps of the main.

A working jib would normally provide good balance to the full unreefed main in wind strengths up to force 4 or 5, that is 11 to 21 knots of wind. To say that a yacht is well balanced implies that the jib, which tends to push the bow away from the wind – that is, to leeward – is matched by the main, which tends to turn the boat to windward.

The Mainsail

The mainsail, or just plain 'main', has identically named parts to the jib, but whereas the luff of the jib is the hypotenuse of its shape (see Fig. 6a), the luff of the main is vertical, and is attached, usually with plastic sliders, to the rear of the mast (see Fig. 6b). The foot of the main fits into a channel on top of the boom. The main is hauled aloft by the main halyard, and is stretched out along the boom by another piece of running rigging called the outhaul. The leech of the main is fitted with a leech cord.

Sail Balance

When sailing in rough weather it is sometimes desirable to reduce the size of the sails, in which case the jib is simply replaced by a smaller one. The main, however, is reduced in size, exposing less of its area to the wind. This is achieved by 'reefing', an operation in which the main halyard is lowered sufficiently for the reefing cringles at tack and clew of the main to be secured to the boom. To complete the reef, the now loose bottom of the sail is bagged along the boom using the main's reefing points; please notice that the main may be reduced to three different sizes by reefing.

A yacht is said to be nicely balanced when the jib, main and tiller are so set that the boat is moving at maximum speed for the wind and sea conditions, and needing only a light pressure on the helm to keep it on course.

The helm provides a convenient indication of sail balance. When the jib and main are nicely balanced, the tiller will be almost in line with the boat and very little pressure will be required by the helmsman to keep it there. I say *almost* in

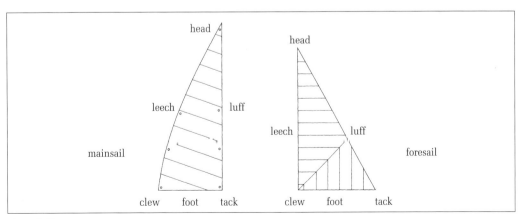

Fig 6 Sails.

line with the boat because it is usually preferred to have a little 'weather-helm': that is, the jib and the tiller, offset slightly to windward, are working together to balance the main.

Where the wind strength is force 4 or less, the skipper may well want to replace the working jib with increasing sizes of genoa; conversely, if the wind speed increases, the skipper will want to reduce the sail area in stages to match the increasing wind strength. The main will be reefed down in stages until, with three reefs in the main and the working jib replaced by the storm jib, the sail area of the yacht is reduced to its minimum. If the wind strength continues to increase, the skipper may decide to heave to.

Wind Speed and Strength

A wind speed of 22 to 27 knots, that is, force 6 or a strong breeze on the Beaufort scale, is often called a small boat gale simply because the sea state under wind strengths of these values makes small boat sailing very uncomfortable, particularly when going to windward. That is not to say dangerous, and an experienced skipper with a good crew and a well-found yacht may well relish such wind in a small boat; however, the elderly and novices should not be taken to sea for the first time in such high winds.

It should be noted that a general forecast wind strength of 4 to 5 may well contain gusts of force 6; also that, having gone out in a wind force of 4 to 5, you may well get caught out in a stronger blow such that the sail area ought to be reduced. The perennial question is 'when to reef': if you are cruising and you think about reefing, then that may well be the time to do it; if you are undecided and the wind strength increases to the extent that the lee rail is awash, you've probably left it too long – certainly the crew on the fore-deck wrestling to change down the jib, or reefing a flogging main on a wet and sloping deck, will think so. If, on the other hand, you are an experienced skipper with a good crew and boat and are racing, any suggestion I give here to reef early would probably be ridiculed.

Rope-Work

The word 'cordage' covers all sizes from thick rope to thin string; it is measured by its circumference, usually in millimetres. Cordage of a circumference greater than about 51mm (2in) is technically rope; below this size it is referred to as 'small stuff'. Historically it is made from natural fibres such as manila, sisal or hemp, nowadays, however, the vast majority of rope on a yacht is of man-made fibres such as nylon, polythene, polyester and kevlar. Nylon is elastic and is therefore ideal for use where sudden loads are experienced, such as anchor warps: a yacht riding to an anchor in heavy seas will tug constantly with considerable force against the restraint of her anchor line, and some elasticity in this line makes for an easier ride. Polyester, under the trade name Terylene, has very little elasticity, making it ideal in situations where tension, once applied, must be maintained until deliberately released, such as in sail halyards; once a sail halyard has been hauled taut for sailing to windward, any easing of the halyard due to unwanted stretch will soften the sail's luff, thus preventing the boat from sailing close-hauled.

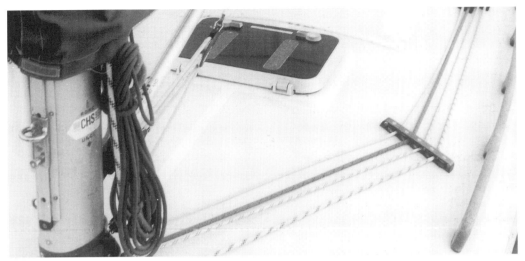

Reefing pennants coiled at mast. Control lines led to cockpit.

Natural Fibre Rope

Natural fibre rope derives its strength from the inter-fibre friction set up between the short fibres, from which these ropes are made, when the rope comes under stress. The fibres, or more strictly the filaments of rope using man-made materials have an inherent strength and extend throughout the length of the rope.

The manufacture of natural fibre rope begins with the twisting together of the short fibres, usually right-handed, into a 'rope yarn'. Rope yarns are twisted together, left-handed, to form a 'rope strand', then three rope strands are twisted together, right-handed, to form the finished traditional 'hawser-laid' rope (Fig. 7). Natural fibre rope must be handled and coiled in a proper manner in order to avoid kinks (twists and loops in a rope which prevent it running freely). Man-made materials are often woven or plaited, producing a relatively kink-free rope.

Natural fibre ropes need much greater care than man-made alternatives, because too sharp a bend can distort and permanently damage them; for example, to avoid damage when passing such a rope through a block, the sheave diameter should be nine times the diameter of the natural rope. Dampness, heat and mildew are all enemies, and such rope should be stored dry, in well ventilated lockers.

Fig 7 Hawser laid rope.

Man-made Ropes

Man-made ropes do not suffer from dampness or mildew, nor do they absorb moisture; they may, however, become hard and brittle after long exposure to sunlight. It is probably true to say that the vast majority of rope-work found on a modern yacht is woven or plaited man-made material. A yacht's rope is so constantly used that its good condition and strength are too easily taken for granted; all cordage should therefore be periodically examined for signs of damage and replaced when signs of wear, chafing or broken fibres/filaments are seen. It must be realized that sheets, halyards, warps, anchor rope and shorelines – in fact, all yacht rope – can be placed under very heavy load, and although designed to take such loads, the use of a damaged piece is asking for trouble; lying to a mooring in a fast-running tide is no time to remember that the mooring line is badly chafed.

Coiling a Rope

A major part of the enjoyment of going to sea in a yacht is the ability to participate in the efficient handling of the boat. A 'sailor' is instantly recognized by the way he/she handles rope-work: the apparent ease with which he coils a rope, and the speed with which a particular knot is completed, indicates a practised skill well worth the learning. A book such as this can only show how a rope is neatly coiled, and how a knot or a bend is efficiently executed; a beginner will do well to obtain a piece of rope and practise all the basic knots shown below.

Assuming that an irresponsible, inexperienced sailor has left a length of laid rope lying on the deck, your immediate task, without waiting to be told, is to coil that rope and stow it in the appropriate locker. First run the offending rope completely through your hands from end to end, undo any knots you find and take out any kinks. To form the coil, take a one-handed hold of the rope about half a metre from one end, and with the other hand take up the rope again about a metre from the held end; form a loop by passing this to the hand already holding the rope. Please note that as you form this first loop, and all subsequent loops, it will be necessary to give a right-hand twist to the rope of about half a turn; with practice,

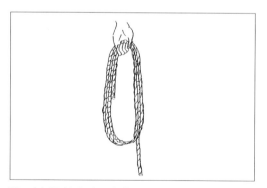

Fig 8(a) Unkinked coiled rope.

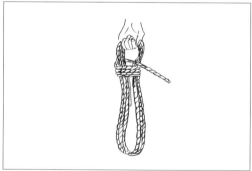

Fig 8(b) Trapping the bitter end to keep the coil secure.

this required twist is applied by rolling the rope between finger and thumb as the coil(s) are formed.

Repeat this loop-forming action until all but one metre of the rope lies in neat coils suspended from your hand (Fig. 8a). Take up the remaining rope and lay it around the suspended coils at a position just below your hand; the coil is completed by trapping the end of the rope as shown in Fig. 8b. Coiling a rope needs practice, and probably a demonstration, so don't be afraid to ask. The 'right-hand twist' mentioned above prevents the length of rope yet to be coiled from kinking. The size of the loops in the coil will depend on the length of rope; with a very long or heavy length of rope, it may be easier to lay the coils onto the deck, only picking it up after completion.

Knots

THE HALF-HITCH

Holding a short length of rope between your hands and twisting one end relative to the other will form a half-hitch. Passing a rope around a bollard also forms a half hitch, and Fig. 9a shows how the friction set up between a rope and a bollard using only a half-hitch can be used to hold, temporarily, a very heavy boat alongside a mooring.

When bringing a yacht alongside a pontoon mooring, the skipper will have organized a bow and stern breast rope, each one brought back to the shrouds and held ready by a member of the crew; as the boat comes alongside, the crew will step, not jump, onto the pontoon. The bow breast rope is taken to a pontoon bollard towards the bow, the stern breast to a bollard astern. Half-hitches are formed round the bollards and the boat brought gently to a halt. Sometimes a skipper may approach the pontoon with too much way on – that is, going too fast as he approaches – and in this circumstance the crew on the pontoon will bring the boat to a gentle stop by 'easing' the half-hitches, allowing the friction to build up as the boat is brought slowly to a halt in a controlled way. If this easing is not carried out, the half-hitches are powerful enough to bring the boat to a sudden stop, crashing against the pontoon in a very unseamanlike manner.

THE CLOVE HITCH

Two half-hitches, made around a spar and flowing in the same direction, form a clove hitch (Fig. 9b). This is a quick and easy knot to tie, but one that can fail in

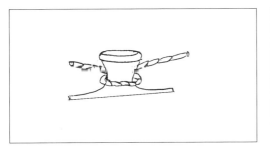

Fig 9(a) Half-hitch.

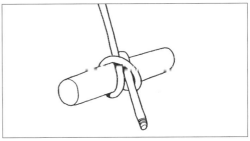

Fig 9(b) Clove hitch.

situations where the load on the hitch is not constant. This hitch is often used to suspend fenders from handrails, and is often the reason that fenders can be seen floating away all by themselves; a more secure way to attach them is to use a round turn and two half-hitches.

A ROUND TURN AND TWO HALF-HITCHES

This is a very secure way of attaching a rope, and unlike some others, it can be undone under load; it is therefore ideal for securing shorelines.

Note: From a study of Fig. 10 it can be seen that the two half-hitches, when tied correctly, form into a clove hitch.

ROLLING HITCH

There are occasions on a yacht when the strain on a rope is such that it cannot be released. One good example – although it is actually an example of bad rope handling – is a 'riding turn' on a winch. In this event that part of the rope which is under tension has ridden up over the turns around the drum and can be very difficult to remove; a rolling hitch just might help to solve the problem by allowing the tension to be eased. As shown in Fig. 11a, it comprises three half-hitches,

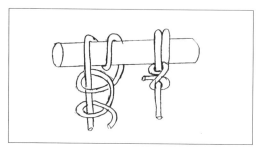

Fig 10 Round turn and two half-hitches.

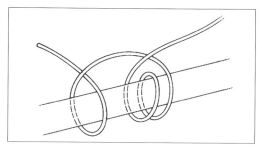

Fig 11(a) Rolling hitch.

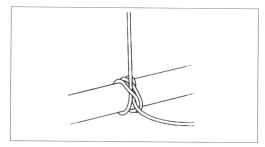

Fig 11(b) Rolling hitch relieving tension.

very similar to the clove hitch – but notice that the second half-hitch is used to 'lock' the first, before the bitter end is tucked away as the third half-hitch. Fig. 11b shows the rolling hitch applied to a rope under tension: pulling in the direction of the arrow will ease the tension on the rope with the riding turn allowing it to be released.

For efficient operation the rope used to make the rolling hitch must have a smaller diameter than the rope under tension.

BOWLINE

The bowline is perhaps the most popular knot on board a yacht; with practice it is quickly made, and provided it is not under tension, it can be undone easily (broken or capsized). Fig. 12 shows how the bowline produces a neat temporary eye at the end of a rope.

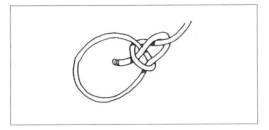

Fig 12 Bowline.

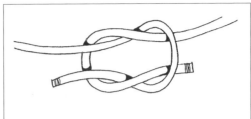

Fig 14 Reef knot.

SHEET BEND

This is used to join two lengths of rope together to form a longer one (Fig. 13a); a double sheet bend is a more secure version (Fig. 13b).

FIGURE-OF-EIGHT KNOT

The figure-of-eight or stopper knot (see Fig. 15) is used at the end of a rope to prevent it from escaping back through, say, a turning block.

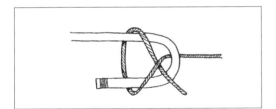

Fig 13(a) Sheet bend.

Fig 15 Figure-of-eight knot.

There are many more useful knots for the sailor to learn to tie; however, the ones detailed above are the most common. Remember you may be dealing with rope-work in poor light and rough seas, so practise all your knots until you can complete them without thought.

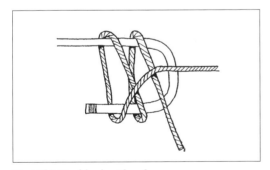

Fig 13(b) Double sheet bend.

REEF KNOT

Originally designed and still used for 'tying in' reef points when shortening sail, the knot is easily undone by pushing the two ropes together, as shown in Fig. 14.

Whipping

A natural fibre rope when cut will quickly unlay and become a frayed, untidy mess unless the end is whipped or back-spliced.

There are three types of whipping: sailor's, common and sailmaker's, and each sort will maintain a tidy bitter end to a rope and still allow it to run freely through a block. Common whipping is not very secure, however, and should be considered as a temporary whipping only. Whipping is carried out using polyester, cotton or flax roping twine; the three methods are shown in Fig. 16.

For sailor's whipping, the end of the twine is laid along the rope, and a number of turns of the twine are made over it, working towards the end of the rope. A bight is made at the working end of the twine with the end facing away from the end of the rope (Fig. 16). Several more turns of twine are laid over the bight until finally, pulling the captured end of the twine, the bight is reduced to zero and the whipping is complete. Common whipping is even simpler to make: see Fig. 16.

Sailmaker's whipping requires a sailmaker's palm and needle; it is, however, the most secure of the three types and, where time permits, should always be employed. The end of the twine is fastened by stitching it through a strand. Several tight turns working against the lay are completed, and then the needle is passed under a strand and brought along the groove between strands; the turns are 'frapped' tight into the groove before being passed under a second strand, and the procedure repeated for all strands. The whipping is completed by stitching the twine through a strand and neatly cutting off.

Splicing

Long, short, eye and back-splicing a rope are terms generally applicable to laid natural fibre ropes and are therefore not now commonly practised. They are, however, worth a mention in any book on seamanship.

THE SHORT SPLICE

Long and short splices are used for joining two lengths of rope together. The short splice is formed as follows: the three strands of each rope to be joined are unlaid for about three times the rope's circumference; the strands of each rope are then married – that is, each strand of both ropes is woven into the other in an 'up, over and tuck' manner. First, a strand of, say, the left-hand rope is passed over the adjacent strand of the right-hand rope,

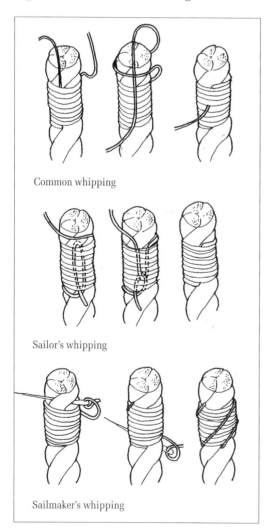

Common whipping

Sailor's whipping

Sailmaker's whipping

Fig 16 Whipping.

and is then tucked under the next strand of the right-hand rope; this operation is repeated using the remaining two strands of the left-hand rope. The three right-hand strands are then married into the left-hand rope using the same up, over and tuck method of working against the lay of the rope. The whole operation is repeated on both ropes until only a small length of each strand remains; this is neatly cut off to complete the short splice. Please note that a short splice adds considerably to the rope's diameter, so that the extended rope may not pass through a block.

THE LONG SPLICE

In a long splice the strands of each rope are unlaid for several inches and then married so that each strand from one rope is paired with a strand of the other; then one strand of a pair is unlaid while its partner replaces it in the lay of the rope. An overhand knot completes the replacing operation, after which an 'over-and-tuck' of the remainder of the two strands finishes off the first two strands . Now a strand is unlaid and replaced in the opposite direction, again ending with an overhand knot and an 'over-and-tuck' to complete. For added strength and neatness, a second over-and-tuck is made on each of the paired strands, but the fibres of each strand are reduced by half before doing so. A long splice does not increase the diameter of the rope except by a small amount at the position of the overhand knots.

THE EYE SPLICE

The eye splice uses the same strand-weaving technique as above; however, a seizing is put on the eye before starting.

THE BACK-SPLICE

This splice is started with a 'crown knot' and completed using the up, over and tuck weaving of the short splice. In fact the back splice is a not strictly a splice because it is not a join of two ropes; it is, however, a neat means of finishing off a cut rope which, without a back splice or a whipping would quickly unlay and become frayed at its end.

Winches and Cleats

Cleats are essentially rope-jamming devices associated with winches and fairleads; they should be so fitted that they lie at an angle to the lead of the rope, presenting an open and closed approach for the rope being cleated. Fig. 17a shows the correct, unjammable approach of a rope coming from a fairlead or from a winch, around which three turns have been made. To cleat a rope properly, a full round turn on the cleat followed by two figure-of-eight turns and a finishing-off second full-round turn, to pull the rope into the cleat, are all that is required (Fig. 17b). It is possible to follow the two figure-of-eight turns with a twisted or 'foul turn'; however, this finish could jam under load, and foul turns, although in common usage, are not recommended in any circumstance.

When the time comes to release a winched and cleated rope, care must be taken to keep a light but necessary hand tension on the rope as it is uncleated; this hand pressure maintains the friction of the three turns around the winch drum. There are two ways to release the winched rope:

(i) Slowly and under full control, as when easing a jib sheet. While keeping sufficient hand tension on the turns

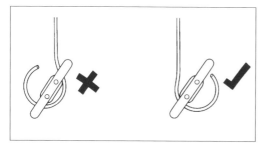

Fig 17(a) Approach to a cleat.

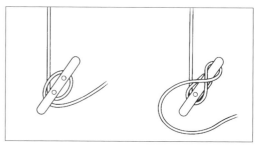

Fig 17(b) Cleated off.

3 good turns round winch, cleated off and coiled.

Warp, properly cleated can not jam.

around the drum, ease the rope with your free hand. It is very important that the hand used to ease the turns on the drum

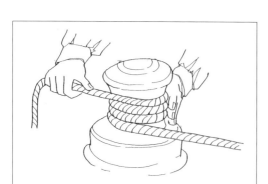

Fig 18 Easing winched turns.

is positioned as shown in Fig. 18; also, take great care that this hand cannot be drawn into the turns on the drum – and if you are unsure, *Ask*!

(ii) Quickly but safely, as when going about. The working jib sheet becomes the lazy sheet and must be released very fast. Uncleat the working sheet, remembering to keep sufficient hand tension to support the winch friction on the three turns. If possible, stand over the winch, and when the command is given, flick the turns off the top of the winch; ensure that the sheet can run freely, and make sure you are not in the way.

2
SAILS AND SAILING

The History of Yachting

The popularity of sailing for pleasure in the UK dates from 1660 when King Charles II was presented with a yacht recently built for the Dutch East India Company. She was called *Mary*, was 52ft in length, with a 19ft beam and 3ft draught, and was originally sloop-rigged with a sprit for the mainsail. She was re-rigged in 1662 with the more popular gaff mainsail. Unlike most privately owned yachts of today she also carried a crew of twenty and ten cannon; however, considering that smuggling and indeed piracy were rife, safety at sea in those days probably depended more on the cannon than on sail trim.

Shortly after the gift of the *Mary* by the Dutch, an English boat-builder produced two yachts, one each for the king and his brother. Based on the *Mary*'s design, these vessels were single-masted with two headsails, a gaff mainsail, and a square topsail over the main.

The earliest recorded sailing club, dating back to at least 1720, was the Water Club of Cork in south-west Ireland. Originally limited to twenty-five members with an annually elected Admiral, the club met for formal dinners and what we would now call rallies. This club has had a somewhat broken record of existence; however, in 1828 it took the title of Cork Yacht Club, becoming the Royal Cork Yacht Club in 1833.

Almost a hundred years later, in 1815, a group of forty-two English gentlemen, nineteen of whom were titled members of the aristocracy, met in London to form 'The Yacht Club', the first organization in the world to have this title. The club is now the Royal Yacht Squadron, so named by King William IV in 1833 and I believe still the only sailing organization permitted to fly the white ensign. It is also the oldest club in the world without a break in its history.

Modern Sailing Rigs

Modern yachting, dating from the mid-nineteenth century, led to a rapid development in fore-and-aft rigging. Fore-and-aft sails, unlike the square-rig ones of earlier years, take the wind on alternate sides as the boat is tacked through the wind onto a new heading. The most common type of modern yacht carries a single mast with two sails, one ahead of and one behind the mast. The foresail, jib or staysail are all

Beautiful all-wood sailing yacht.

Gardener to his design. A Bermudan-rigged sloop was not only much faster to windward than any previous sail plan, but more importantly, she could sail much closer to the wind.

Sail Trim

A sail's basic shape is built into the fabric by the sailmaker, and it is no bad thing to discuss with him how the sails should be trimmed.

The Jib.

The trim of the foresail is controlled by:

(1)   The halyard: this rope is used to hoist the jib aloft. The harder the halyard is winched home, the tighter becomes the luff of the sail, and by implication the slacker becomes the leech.

names used to define a triangular sail fitted ahead of the mast, the original purpose of which was probably to assist in putting the head of the boat through the wind when going about. Invented and designed by the Dutch in the sixteenth century, this sail, laced to the forestay and hauled aloft on its own halyard, brought a great improvement in boat manoeuvrability. The advent of the foresail was probably the single most important development in wind-powered propulsion.

It was not until 1910 in the Bermudan office of one William Gardener, an American yacht architect, that a triangular mainsail was first set onto a one-piece mast. This rig became known as the Bermudan sloop. The aerodynamic properties of sails were beginning to be studied and understood at this time, and it was perhaps this new knowledge that led

Neatly rolled away furling jib.

(2) The sheet: the tension of the sheet determines how flat the jib sets, and the harder the sheet is winched in, the flatter the jib is set. The angle at which the sheet leaves the clew of the foresail is determined by the position of the car; this angle is normally set to bisect the included angle of the sail at the clew point. Moving the car position forwards from its normal position will increase the tension of the leech and flatten it; the tension along the foot of the sail and the luff will ease. Moving the car further aft will increase the foot tension and reduce the tension of the leech.

(3) A Cunningham cringle is provided just above the tack, this facilitates extra tension for the luff. A leech cord is run down the leech of the sail to quieten leech flutter in light winds.

The Main

(1) The halyard hauls the main aloft and sets the tension of the luff.

(2) The mainsheet in conjunction with the traveller and boom-vang are used to trim the mainsail: that is, to set the main to the correct angle to the wind and to the correct shape and sail tension.

(3) The out-haul is a rope attached to the clew of the mainsail, and it is used to set the tension of the foot of the sail. By moving the clew further towards the outer end of the boom, the tension and flatness of the foot of the sail are increased.

Points of Sail

Sailing to Windward

Fig. 19 shows a yacht moored on a long warp in a tideless water with her jib and main hoisted; she is lying head to wind. In a 10 to 16 knot blow her sails will be flapping rather like a flag atop a flagpole, and for those people on board her, the sails will be making quite a din; the boat, lying dead in the water, will be straining

Neat use of lazy jacks to catch dropped main sail. Zip on bag for easy storage.

against the warp. This situation is not a pleasant one and has been proposed here only to allow the following scenario:

Assume now that the engine is started and engaged, ticking over with just sufficient power to keep the boat's head to wind. If the starboard jib sheet is hauled in now, also the main sheet hardened in to bring the boom amidships, the tiller put gently but increasingly across to her port side, and the engine power only slightly increased, the sails will continue to flap but the boat will move slowly ahead and to the right – that is, she bears off to starboard under the gentle influence of the engine and tiller. This movement to starboard will continue until the yacht is lying at approximately 45 degrees to the wind. At this moment, if the mooring warp is let go and the engine turned off, several magical effects will be noticed:

(1) Both sails will fill with wind and become quiet, taking on the shape shown at Fig. 20.

(2) The boat will lift up in the water and move forwards as she begins to sail into the wind.

(3) The boat will heel over away from the wind: that is, she will lean to leeward.

(4) The tiller will come alive and the helm will need a firmer grip to keep the boat on its 45 degree heading.

(5) As the boat picks up speed under the influence of the sails alone the wind will appear to move forwards, and, from being at 45 degrees to the port side of the boat, it will feel, as it blows on your face, to be almost dead ahead.

(6) There will be the sound of water lapping against the hull as she cuts through the water.

This experience of sailing a yacht into the wind, or beating, is one of life's finest, and it is a great thrill to feel her respond to your movements on the tiller: if you move the tiller gently and lightly to starboard the boat will slow, her sails will flap and she'll sit up straight in the water; moving the tiller back to port will cause her to heel and sail again.

The technical term for the above is that the yacht is beating to windward. To 'point' is to sail as close as possible to a wind of this strength, and in order to do this the jib sheet should be hauled or winched in, until the back two-thirds of the jib is as flat as possible; the foot of the jib will then be almost touching the shrouds. The main sheet should be hauled

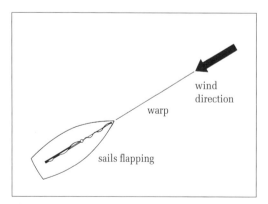

Fig 19 Lying to a warp.

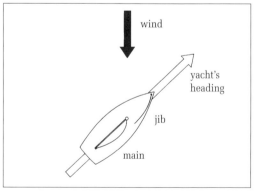

Fig 20 Sail shape.

in to get the boom amidships. Later in this chapter we shall discuss how a sail's halyard, sheets, outhaul and leech cord may be adjusted to give the highest possible point to windward when beating.

In the instance above we hauled in the jib's starboard sheet and moved the tiller across to port, and the boat, at 45 degrees to the wind, began to sail. We could just as well have hauled in the jib's port sheet and put the tiller across to starboard, and the boat would have sailed away to port as she turned at approximately 45 degrees to the wind coming in on her starboard side. With the wind coming over her starboard side she is said to be on a starboard tack; with the wind coming over her port side, she is on a port tack. Any attempt to sail closer to the wind than 45 degrees will fail, the sails will flap, the boat slow down and she will sit up straight in the water. Fig. 21 illustrates the 'no go' area of 90 degrees, centred on the direction of the wind, in which it is impossible for a yacht to sail.

We shall see later that a vessel can sail on all other points of the compass.

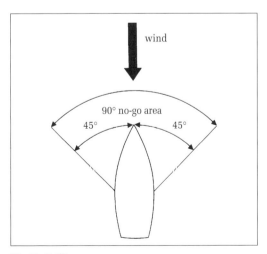

Fig 21 Sailing no-go areas.

Tacking

When a sailing boat needs to make to windward, this is achieved by a series of tacks, first sailing 45 degrees off the wind on, say, a starboard tack, with the port jib sheet hauled in taut and the main boom centred; then by putting the tiller down to the lee side, in this case putting the tiller to port, the boat's head will come up to windward, the sails will flap, the boat will sit up and as the yacht's head moves through the direction of the wind the port jib sheet is let go and the starboard jib sheet hauled tight. Both jib and main will fill at the magic 45-degree point and the boat will be on a port tack.

This manoeuvre is called going about, or tacking, and should be performed as a smooth, quick operation (see Fig. 22). The yacht needs to be moving quite quickly before a tack is attempted, otherwise she may come to rest with her head to wind and stopped; she is then said to be 'in irons', and it may be necessary to hold out the clew of the jib to windward to make her pay off as the wind literally blows the bows round.

When sailing close to the wind the sheets will be hauled tight, the airfoil combination of jib and main must be presented to the wind at the correct attack angle, the yacht will be heeled to leeward, and the exhilaration of speed will be wonderful. When beating it is the helm's responsibility to keep the boat sailing as close to the wind as possible to maintain best 'make' over the ground: too close to windward will cause the sails to flap, and the yacht will sit up and slow down; too far off the wind and the yacht will heel excessively and also slow. Sailing by the wind is a skill only acquired by practice; however, 'tell-tales' can be very

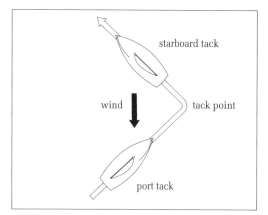

Fig 22 Making to windward.

helpful in keeping best course to windward.

Tell-Tales

Tell-tales are ribbons of lightweight cloth sewn into the luff of the jib and the leech of the main; they give a good indication of when the sails are correctly, or incorrectly trimmed, ie that the sails are at the correct attack angle to the wind. Usually made from spinnaker sailcloth, three tell-tales are fitted to each sail; on the jib they are positioned just above the foot, half-way up the sail and just below the head, and they are fitted to both sides of the jib cloth about 6in (15cm) in from the luff rope. On the main, the tell-tales are sewn directly onto the leech. When the sails are correctly trimmed and the boat is on the correct point of sail, all the tell-tales fly cleanly off the jib sailcloth and stream out straight behind the main.

Other Points of Sail

We have so far discussed how a sailing boat can beat to windward, something

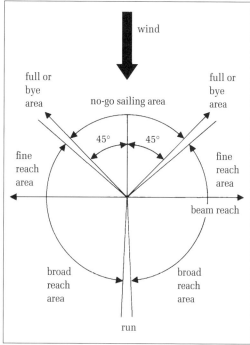

Fig 23 Points of sailing.

which the old square-riggers were quite incapable of doing. When beating as closely as possible to the wind in a good breeze it is necessary for the sheets of both sails to be winched in very hard in order that the sail's airfoil section is properly set and can be presented to the wind at the correct attack angle. It was stated earlier that a sailing vessel will sail on all points of the compass except the 90 degree area centred on the wind. These different points of sail are named and shown in Fig. 23.

Reaching

When a yacht beating to windward bears away from the wind she moves first into the 'full and by' point of sail; continuing to bear further away from the wind, she moves through a fine, beam and broad Reach until, with her stern to the wind, she is said to be 'running before the wind' or just plain running.

In all sailing positions except running, the sails must present the same attack angle to the wind as when the boat is beating. This means that as she bears away from the wind onto a reach, the sheets of both jib and main must be eased in order for the two sails to maintain their essential air foil lift. Dealing with the jib first: easing the jib too much causes the windward tell-tales to lift and flutter; if the easing is continued, the luff of the jib will lift, and eventually will flap. Where the jib is not eased sufficiently, the lee-side jib tell-tales will lift and flutter. Thus the golden rule when reaching is to sheet towards a lifting tell-tale, ie ease to a lifting leeward tell-tale, and harden into a lifting windward tell-tale. When beating, the rule is to luff to a lifting leeward tell-tale, and bear away to a lifting windward tell-tale.

The general rule for the main when reaching is that the sheet is eased until the luff just lifts, at which point the sheet is hardened in just slightly.

Setting the jib and main of a yacht correctly for reaching as described above should be viewed as a good general sailing technique. There is always discussion over precisely how the two sails should be rigged, tensioned and set to make a yacht sail efficiently. In short, the jib is correctly sheeted when both inner and outer tell-tales fly horizontally off the sailcloth. In many reaching situations – ideally force 3 to 4 wind, nice calm sea, sunny weather and a long sail on one tack – the skipper has time to play with the many sail-controlling ropes, and this can be very instructive, especially if the yacht's speed through the water can be monitored.

The Apparent Wind

We have noted that as the boat begins to sail, the wind at 45 degrees to the boat appears to move forwards as though now blowing from almost dead ahead. This apparent forward shift in the wind direction, which increases as the yacht's speed increases, is only experienced by those on the boat – a person walking on the water just a few feet away would not experience any change in the wind direction.

The wind shift noticed by the crew is called 'apparent wind' and is a combination of the wind caused by the yacht's movement through the air as she picks up speed, and the natural or true wind of 45 degrees blowing onto her. The wind experienced by the water-walker is the 'true wind'.

Fig. 24a shows a yacht making 5 knots, through still water, beating at 45 degrees to windward against a 15-knot true wind.

The yacht is under the influence of two winds:

(1) 5 knots of wind from dead ahead due to her movement through the air; to explain this, imagine a calm day with no wind. A boat begins motoring ahead, and as she picks up speed the crew would feel a breeze from dead ahead exactly equal to the boat's speed. Thus motoring ahead at 5 knots would create an onboard 5-knot wind from dead ahead.

(2) The true wind blowing onto the yacht; in the case of Fig. 24a, this is 15 knots at 45 degrees to her heading.

To the crew, these two winds would combine as one wind, the strength and direction of which can be determined by drawing a parallelogram as shown at

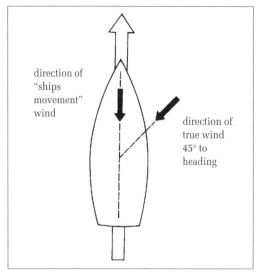

Fig 24(a) Two winds.

Fig. 24b. This figure is scaled: 2cm to 5 knots and the 45-degree angle is accurate, and so we can measure straight from the figure that the apparent wind speed is approximately 19 knots and is 35 degrees from the yacht's heading.

Running

When running downwind the boom is eased out until it almost touches the shrouds, and the mainsail will be brought against the shrouds by this action; that area of sailcloth which will be pushing against the end of the crosstree is usually protected by a sacrificial wear patch.

The jib sheet is eased until the clew and tack are in line at about 90 degrees to the wind; unfortunately, in this position the jib is almost entirely screened from the wind by the main, causing the jib to hang loose and flap. The solution is to goosewing the jib, that is to haul in the lazy sheet whilst easing the working sheet, so that the jib is pulled out to the side of the yacht opposite the main. It is possible, although difficult and improbable, that the goosewinged jib will fill and fly without further assistance, simply under the control of her sheet. However the goosed jib will be much easier to control if it is 'poled out'.

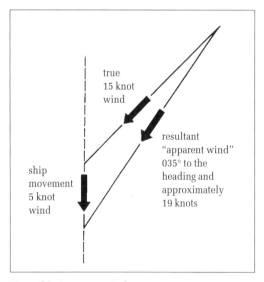

Fig 24(b) Apparent wind vector.

Goose winged.

Running dead downwind can be a difficult and dangerous point of sail, particularly to the inexperienced sailor, for the following reasons:

(1) When running in an increasing wind, the yacht's speed can build up quite quickly almost without the crew noticing; they may find that suddenly a big sea has developed and the boat is surfing along at high speed. The risk of her burying her bow as she runs down a large wave, with the consequent possibility of a broach and knockdown, increases with increasing wind strength and wave height.

(2) The helm must be experienced, since the boat must not be allowed to move into 'sailing by the lee': that is, the wind must not be allowed to get behind the main because this could result in an unwanted, unexpected, uncontrolled and dangerous jibe. This is when the uncontrolled, back-winded main carries the boom suddenly and at high speed across the cockpit, taking off anyone's head in its path before tearing away the opposite side shrouds. The loose main sheet will come across the coach roof sweeping all before it over the side.

I hope you will therefore agree that sailing by the lee on a full downwind run is to be avoided, especially in a following sea when the boat may be yawing quite badly. In fact most of the above risks can be considerably reduced by sailing 10 degrees off a full 180-degree wind, and jibing round as necessary so that the average course is the one desired (see Fig. 25).

Preparing for Heavy Weather

Sooner or later you are going to run into foul weather. The more you sail, the more you will want to sail, and over the years the odds of being caught out in a heavy blow will reduce to the point at which it becomes very likely that you'll be out in strong winds. A well founded boat, an experienced crew and an observant well prepared skipper have little to fear from 'normal' bad weather in UK waters. A force 6 wind, 22 to 27 knots of wind is usually considered to be a small boat gale, and, until your own experience and wise council tell you otherwise, this is not a bad marker for indicating that you should be heading for safe shelter; remember, a force 6 forecast may well mean actual gusts of force 7 or more. It would be prudent to remain tied alongside if such a force is predicted; however, you may already be out there, heading home, when the weather takes a turn for the worse. Early acceptance of the likeli-

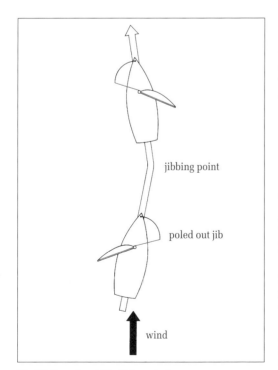

jibbing point

poled out jib

wind

Fig 25 Running 10° off the wind.

hood of strong winds allows sensible pre-cautions to be made:

(1) A reduction of sail area whilst this is still easy to do: change down the jib, reef down the main.

(2) Put on, or at least make ready, an appropriate number of layers of warm dry clothing and foul weather kit. If you have not already done so, demonstrate and individually fit lifejackets (even if this was done in your skipper's pre-sail safety chat, a reminder at this stage – par-ticularly to new sailors – will put crew members' minds at rest and give them something to concentrate on). Fit safety harnesses and secure points of anchorage and jackstays.

(3) Stow things away down below, and check that hatches are securely closed.

(4) Provide food for everybody now: a well fed, warm, dry sailor is better able to stand a bit of rough weather than one who is not. Prepare vacuum-flasks of hot soup and sandwiches.

Many good, comprehensive books and videos are available with such titles as *Heavy Weather Sailing*: read and/or watch them, several times if possible, and take note of what they say. Good indica-tions that the wind has increased to the point at which sail must be reduced are:

(1) The yacht is increasingly heeling to leeward.

(2) A stronger grip on the tiller is needed and the tiller must be held more off-centre, ie further to windward, to maintain the same course. This increased tiller angle also means that the rudder blade is so far off the centreline of the yacht that it is act-ing like a brake on the boat's speed.

(3) Sudden gusts of wind cause the boat to turn up to windward, and in extreme cases the helm simply cannot prevent this

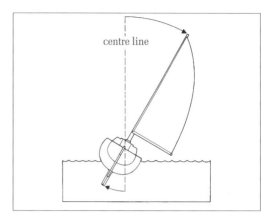

Fig 26 The yacht is well heeled over, the leeward rail awash. The water friction on the underwater hull is much greater on the port side of the centre line than on the starboard side. The yacht is over canvassed, sudden wind gusts will increase the heel angle, the tiller may well be overpowered by the even greater friction on the port side of the hull causing the yacht to pivot uncontrollably to port.

turning motion from happening This is because the underwater profile at steep angles of heel is such that water friction on the underwater windward side of the hull is so much greater than on the under-water leeward side that the tiller loses its power to control the ship's heading. The underwater friction on a heeling yacht is shown at Fig. 26.

(4) In high wind circumstances where the heeling is such that the lee toe-rail is con-stantly underwater, the yacht is over-can-vased and sail should be reduced. This is certainly true when cruising. Even when racing, a permanently submerged lee toe-rail probably indicates an inefficient underwater profile and a braking rudder.

Reefing

Reefing is the act of shortening a sail by furling it and rolling it away into the mast, or into or around the boom. When

Mast winches and reefing pennants.

reefing the main or changing down the jib it is usually necessary for members of the crew to work on the open deck; for this reason the following precautions should be observed:

(1) The operation of reefing the main and changing the headsail is practised in calm seas before such action becomes necessary in bad weather.

(2) The actions, when necessary, are carried out at the first signs of bad weather. When contemplating putting to sea in a force 4 to 5 wind, it is probably wise to tuck in a reef and bend on a working jib before leaving.

When sailing, the weight of the boom is taken up by the mainsail, and as we know, this sail has three controlling lines: the halyard, the boom-vang and the main sheet. A fourth rope is used to support the boom at those times when the main is unable to do so; it runs from the end of the boom, up through a pulley at the top of the mast, down the inside of the mast,

End of boom showing:

1. Leech cord of main.
2. Outhaul and two rigged reefing pennants.
3. Topping lift.

and re-emerges at about waist height to be cleated off. It is called the 'topping lift', and takes the weight of the boom at those times when the sail is to be reefed, or in harbour when the main is stowed and the boom needs to be lifted clear of people moving around the cockpit.

When reefing the main, the wind in this sail must be reduced so that the halyard can be eased, thus allowing a crew member at the mast to haul down on the luff of the sail until the appropriate reefing cringle in the sail can be hooked over the 'ram's horn' fitting; the associated reefing pennant is then hauled taught, often with the assistance of a mast winch, to pull the foot of the reduced sail out along the boom to the correct degree of tension. The 'loose bag' of the now unused bottom of the sail is gathered neatly to the boom by means of the 'reefing points'. Please note that the reefing points are to be used only to 'gather' the sail to the boom; actual tension on these reefing points should be minimal.

To reduce the wind in the main prior to reefing, the skipper will either cause the yacht to head up almost into the wind, or cause the yacht to 'heave-to'. If the first method is chosen, the jib will almost certainly flap around and in doing so will interfere with the crew at the mast, the jib clew with its attached 'bowlined' sheets becoming a missile of some considerable force. If time permits and when cruising – *not* racing – heaving-to is a much quieter and more controlled method of preparing to reef the main. Whichever wind-reducing method is used, the first action is to take the weight of the boom onto the topping lift. The main can then be reefed, and when the operation is complete the yacht sails away with the topping lift eased as

the sail again takes the weight of the boom.

It is worth mentioning that if the weather worsens and the skipper leaves reefing until the lee rail is in the water, then heaving-to *should* be used in preparation for the reef.

Heaving-To

Heaving-to is a method of sailing which causes the yacht to move at some 60 degrees to the wind at a speed of less than one knot; as above, it can be useful when reefing the main in heavy weather. It can also give the crew a respite from beating in heavy weather sailing, since the boat will sit up straight in these circumstances allowing food and drink to be taken. It is also a pleasant way for the entire crew to take afternoon tea in the cockpit on a long cruise.

To begin heaving-to, the yacht is brought onto a beat, followed by the helm putting the yacht's head through the wind – *but* unlike a tack, the working jib sheet is *not let go* when heaving-to. Consequently the yacht bears away on a new heading, and with the jib 'backed' the helm must be reversed as the head goes through the wind, and at the same time the main sheet eased so that the mainsail flaps in the wind. At this time the mainsail is ineffective, providing no power to drive the boat. It is the backed jib which is driving and pivoting the boat to leeward, and as the yacht gathers way on this new heading, the rudder will bite and in doing so will pivot the yacht to windward; consequently the jib will lose drive, the boat will slow, and the rudder will cease to be effective. The yacht will then fall away to leeward, and again gather way.

The backed jib and rudder are working in opposition, and only when the boat picks up speed will the rudder provide a turning moment; their combined effect is for the yacht to 'scallop' across the wind at slow speed. In Fig. 31 the backed jib is pivoting the boat to starboard and providing drive. The fully reversed tiller, only effective when the boat has gathered way under the influence of the jib, will turn the yacht to port, the jib will lose drive, and the boat will slow.

The yacht may take some time to settle down, but once she does so, the tiller can be lashed down – and please note, it should be 'lashed down' in such a way that it is ready for instant release if required. The main is quietened by slightly taking in the main sheet, and in some cases the main boom can be hauled into the centre of the yacht without upsetting the balance of the jib/tiller combina-

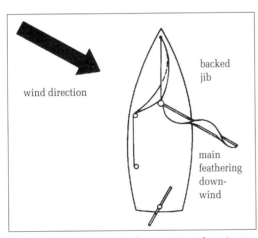

Fig 27 Hove-to seen from above; prior to heaving-to this yacht had been on starboard tack. To heave-to the helm had put the tiller over to port, keeping the port sheet under tension as the bow came through the wind has 'backed the jib'. Quickly reversing the tiller to the position shown completed the heave-to operation. The yacht will now 'sail at some 60° to the wind at less than one knot.

tion. As with most operations at sea, heaving-to should be practised in fine weather.

Bare Poles

In developing bad weather the prudent skipper, having reduced the sails to a minimum and with perhaps a spell of heaving-to as the weather worsens, may find that even under the minimum of canvas the yacht is moving too fast through the water. The last resort is to take off all canvas and either lie a-hull to a sea anchor, or to run before the wind under bare poles.

In lying a-hull, with all sails removed and stowed below, the tiller is lashed to leeward and the yacht will take up her own position with regard to the wind and sea; the rolling motion of this position is more uncomfortable than dangerous. Lying a-hull to a sea anchor and under bare poles will slow the boat's progress to a minimum; however, the sea anchor can be a brutal brake causing massive strain to the boat, and it should only be used when sea-room is reducing, as for example on a lee shore. The problem with lying a-hull is that in a confused sea a freak large wave may throw the boat onto her side, causing damage to the superstructure. As in all things at sea, the skipper must decide how to respond to a particular heavy weather state, and if when lying a-hull the seas continue to mount to dangerous heights and with confused patterns, the decision to run before the storm may well have to be made.

When running under bare poles the boat will have considerable windage due to her freeboard, masts and rigging – in other words, she can be steered, and in fact she may well still be travelling too fast under these condition. In this case, warps they may be streamed astern to slow her down; may be in the form of a long bight 100m to 200m long, or several single ropes may be employed. The warps should be constantly under water to stop then snatching at the yacht, and if necessary weighted so as to maintain a constant pull.

In some cases of strong winds with a long fetch, the following seas may well poop the yacht – that is, the cockpit will fill with water as the sea breaks over the stern. Essentially therefore, the companionway weather-boards must be *in situ*, and the cockpit drains kept clear.

Boat Handling

Boat handling covers such manouvres as letting go and picking up an anchor or mooring buoy, coming alongside, and letting go from a floating pontoon or harbour wall. It also embraces the ability to move around a small, crowded harbour under sail or under engine at slow speed without endangering or upsetting other owners and their boats.

Boat handling, and especially in close quarters to other boats, requires a knowledge of how a combination of tidal stream flow and wind will affect your boat, and of how your boat responds to prop-walk, windage and leeway, and of her behaviour when going astern. Some of these subjects are covered elsewhere in this book, but only in theory; however, if you are unfamiliar with them, please read them before continuing here.

It is only common sense to practise all the boat-handling operations referred to below in calm, safe, uncrowded locations, and to practise them often. It is also very tempting to write here that anchoring, mooring and alongside manoevring should all be done under engine; however, engines do fail, they also occasionally refuse to start, and plastic things do get themselves tightly wrapped around props. Inevitably this will happen just as you've entered an unknown, small and

When not sailing, but driven by engine – fly the black triangle.

crowded harbour, also just after you've dropped the bagged jib down the hatch and put the cover over the neatly furled main. It follows that if you do intend to manoeuvre in close proximity to others under engine alone, then your sails must be available for a fast hoist.

It is important to use tidal flow to your advantage. Just to remind you, tidal flow may be with you, thus increasing your speed over the ground; against you, thus acting as a brake to slow your speed over the ground; or it may be edging you off your intended direction. This last can be determined by looking ahead, in so far as an offsetting stream will cause you to 'crab' away from your heading. However, you can use it to help you: for example, always use the stream or current as a brake by heading into it when coming alongside. One knot of current is equivalent in braking power to a force 3 or 4 wind, and a 2-knot tide is as effective as a force 5 or 6 knot wind.

The wind is another important factor to consider, and you should determine which is the stronger, wind or tide. Or maybe wind and tide are acting together? Or is the wind flowing against the tide? Or is it in fact a cross-tide wind? Study the area as you approach, and if in any doubt about the correctness of your assessment, go round and try again.

Anchoring

There are many reasons for wanting to lie to an anchor: sheltering in a bay from a storm is an obvious one, and waiting for the tide to turn in your favour is another; a small yacht sailing through the water at some 4 knots will not cover much ground against your average foul tide, so you may as well anchor for a pleasant swim or for

At anchor; trailing rope is a long dinghy painter.

tea and scones in the cockpit with all the crew. Whatever the reason for doing so, anchoring a yacht requires planning and the following contingencies should be considered:

1) The depth of water required: the skipper will know the draught of the yacht, and will be able to calculate how the depth of water will change during the anchoring period.

2) The skipper will also recognize the possible danger of anchoring off a lee shore.

3) The approach: in an empty anchorage with no other boats to consider, the yacht is brought to the desired depth, at which point the anchor is let go. In busy anchorages a judgement will have to be made regarding the swinging scope of yachts already anchored, and the relative strength and direction of wind and tidal flow. Choose, if you can, the leeside of similar yachts using similar ground tackle.

Fig. 28 shows the most popular yacht anchors; although some yachtsmen might baulk at carrying the older fisherman's anchor, it is still seen, particularly on more traditional boats.

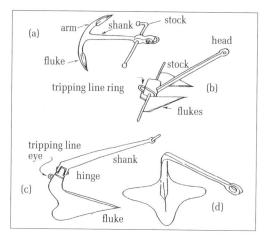

Neatly stowed Bower anchor. Note the samson post on the fore-deck. Very strong point of attachment.

Fig 28 Various Anchors.

a. The fisherman's anchor stows flat, has a low holding to weight ratio.

b. The Dartforth anchor stows flat, has a good holding to weight ratio.

c. The Plough/C.Q.R has a high holding to weight ratio, is awkward to handle and difficult to stow

d. The Bruce anchor has a high holding to weight ratio, is bulky to stow and is often seen leaning nonchalantly over the bow roller.

THE MINIMUM LENGTH OF CHAIN

All anchors should be attached to at least a minimum length of chain, 'minimum' in this case being approximately 25 per cent of the anchor's 'normal' cable length. I hesitate as I write this, because how do we asses an anchor's normal cable length? Since the length used at any one time depends on the depth of the water and the sea state, we cannot fix this 25 per cent. However, once the anchor is down and holding, it is very desirable that any pull is in a horizontal direction so that it causes the anchor to be more firmly pulled into the sea bed. This is why a length of chain is attached to the anchor initially, it's weight causing it to lie horizontally along the sea bed. In all cases of anchoring, sufficient length of cable is deployed to guaran-

tee that the yacht is held by the catenary curve of the cable, and that it does not pull directly on the anchor.

In calm weather, and assuming that the whole anchor cable is chain, then the cable length should be four times the maximum depth of water to allow sufficient scope for safe anchoring. However, if the anchor cable consists of minimum chain and then warp such as nylon, then a length of six times the maximum depth is recommended (see Fig. 29). Let us now discuss the length of 'minimum' chain.

Sailing in the Solent, the maximum anchoring depth might be 5m and so this

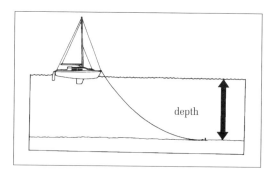

Fig 29 Cable length = 4Xdepth for chain, 6Xdepth for warp. Where depth means maximum depth, sea state is calm, current is slight and wind is forecast to remain low strength.

would require, by the formula, 30m of anchor cable; therefore 7.5m of chain would be quite sufficient. *But* you might take the boat out of the Solent, you may be anchoring in far-from-calm seas in greater-than-expected depths of water – and you may want to sleep peacefully while doing so.

Fit chain, therefore, for the whole of the anchor cable length; after all, what are the disadvantages of doing this? Only money and the extra weight forward – and, oh yes! what you put down, you have to pull back up again.

In bad weather the yacht pitching and snatching at the anchor will require more scope: 'If in doubt let more out'. This is all very well if you have plenty of sea room, and not too much company in the way of other boats at the anchorage; each anchoring circumstance is unique, and once again it is the skipper's responsibility to anchor safely in the prevailing conditions. Note, too, that the amount of chain deployed in fine weather in a calm sea at tea-time may well not be enough at 2am in worsening weather.

PRACTICAL TECHNIQUE OF ANCHORING

In preparing to anchor, the full length of chain required is hauled out of the anchor locker, and flaked out on deck by the foredeck crew with the end of the required length secured to the samson post or some other strong foredeck fitting; the anchor is made ready for lowering over the bow.

Assuming that the yacht is motoring, the skipper will manoevre to the desired position, either up-tide or motoring head-to-wind, depending on their relative strength. At the chosen position, stopped in the water, he will call: 'Let go when

Jib stowed, Anchor flaked out on deck

ready'. The anchor is then lowered, hand-over-hand, until it is felt to be on the bottom; the skipper is informed of this, and he will put the yacht gently into reverse or will allow it to fall away under tide or wind, thus causing the remaining chain to be pulled through the crew's hands. When all the required chain is out, the skipper is again informed; the yacht continues in reverse motion, however, to 'bed the anchor into the sea bed'.

The above operation requires full communication between the foredeck crew and the helm; it also requires that the anchor cable is free to run. The foredeck crew should wear good strong gloves. Once the anchor is down, the skipper must check that it is holding and not simply dragging across the sea bed; note that the yacht may well take some little while to settle. How she lies will depend on the relative strength and direction of the tide and wind. When approaching the anchorage the skipper will have observed how yachts already at anchor were lying, because this will have helped to determine his approach. Compass bearings of shore objects or convenient transits will determine whether or not the anchor is holding, and these should be checked during the anchoring period. One other good method of checking the anchor is to grip the anchor chain firmly through the pulpit: if the anchor is dragging the chain will 'grumble' in your hand.

ANCHORING IN A POPULAR ANCHORAGE

In this circumstance the skipper should note the ground tackle of the various yachts already lying to an anchor, and always anchor near boats using similar cable, ie chain near chain, warp near warp. This is necessary because ground tackle quite often determines how a yacht lies to her anchor, and it certainly dictates how quickly she swings round in sympathy to a changing tide direction.

Strictly speaking, an anchor watch should be maintained, but since this is not always possible, the skipper must ensure the security of his anchor and the safety margin of his own and adjacent yachts should they swing circles, *before* going ashore.

PICKING UP THE ANCHOR

The skipper motors towards the anchor very slowly, giving the crew time to haul the cable on deck, and if possible to feed it into the chain locker at the same time. At the point where the cable is hanging straight down into the water, the informed skipper will stop the boat, and the foredeck crew will continue to heave until the anchor breaks through the water surface. The skipper can then bear away as the anchor is brought onboard and stowed.

Anchoring under Sail

If the anchoring position can be approached on a fine reach it may be better to come up under mainsail alone, with the jib stowed, because this allows much more freedom of movement for the foredeck crew. The yacht's approach speed is controlled by easing the main sheet, and in most cases it can be almost stopped in the water at the desired point with the main fully eased and feathering

All at anchor.

downwind. The anchor should be let go smartly, and, once it is holding, the main should be dropped as soon as possible to prevent the boat sailing around the anchor.

In strong wind and tide conditions the main could prove to be quite a problem in this manoevre, especially while waiting for assurance that the anchor is holding; even so; you cannot afford to drop the main too soon in case you need to sail away and try again. For this reason, anchoring under sail is something to be well practised, and is not recommended in crowded anchorages.

Where the approach to the anchor point is such that the wind cannot be spilled from the main – that is, on a beam reach or run point of sail – the main must be dropped before the approach is made. Proceed under jib alone, being prepared to drop this sail, or have it only partially hoisted if travelling too fast; or in strong winds, approach under bare poles. In any case, drop the jib on the final approach to clear the foredeck.

WEIGHING ANCHOR

To weigh anchor under sail with tide and wind together, first hoist the main – you may have to haul the boom to windward to get the boat sailing away on one tack or the other – and take in the slack chain as the yacht moves; however, be extremely aware that the yacht is sailing at some 45 degrees away from the anchor point, and that as she reaches the end of her scope a very strong tug will haul her round onto the other tack. You must protect your fingers. Haul in the slack chain with a round turn around the samson post or similar strongpoint, keep a sharp eye on the incoming chain, and cease hauling long

before the 'tug' comes on. It may take several forced tacks before the anchor is broken out, but they will be safe tacks, completed by a foredeck crew with fingers still intact.

Tripping Lines

All anchors have a tripping eye on the shank, behind the flukes, to facilitate breaking the anchor out backwards. A non-buoyant line secured to this point and held at sea level by means of a small float may enable the crew to heave a 'stuck anchor' out from under an obstruction such as other ground tackle or one fouled in sea-bed rock. At least, that's the theory, and certainly if you don't fit a tripping line, a stuck anchor is likely to stay stuck and may even have to be abandoned – and abandoning your anchor is expensive and leaves you vulnerable.

Anchoring in Strong Winds, Tides or Currents

You can increase your holding power by anchoring fore and aft: first, lay the bower anchor upstream of where you want to lie, then drop back to approximately double the length of cable you want to lay; lower your second anchor, then motor up towards your final position, take in the fore cable while at the same time letting out the stern cable until the desired midpoint is reached. If there is a strong tide the biggest anchor should take the greater strain.

A second option is to ride to two bow anchors, and for maximum benefit the anchors should be about 30 degrees apart. Lay the first anchor, and, whilst paying out its anchor chain, motor off to one side to let go the second. Finally position the boat, completing the operation by shackling the second anchor line to the first.

Mooring Buoys

Which is the stronger, wind or tide? If possible, see how boats already at a mooring are lying, because their angle is probably the best approach angle for you to try. Picking up a permanent mooring under engine is similar to anchoring in that the skipper will gently motor up to the mooring buoy with just enough way on to keep steerage; the fore-deck crew will then neatly pass a full turn of the bitter end of a secured mooring warp through the top eye on the buoy as the skipper brings the yacht alongside it; finally the bitter end is brought back on board to secure on the fore-deck.

In many cases a small float is permanently attached to the eye of the mooring buoy by means of a short rope. This small float is captured using the boat-hook, brought on board and secured on the fore-deck.

Mooring under Sail

This manoevre *definitely* requires practice, and yet more practice. Let us assume first that wind and tide are together and the tide not too strong. An approach on a beam-to-fine reach will enable you to make against the tide, spilling wind to control your speed over the final few metres. Picking the mooring up alongside, rather than dead ahead, gives the helm a sight of the buoy right up to the last moment.

When approaching the mooring buoy, a transit must be made between the buoy and some object, and the approach to the buoy is made along this transit. In these same wind/stream conditions, a strong tide makes this approach very difficult, the problem being that you must make against the tide towards the buoy and the approach line is a very difficult one to judge; it may even be that a strong tide will defeat all your attempts.

An alternative method is to approach downwind and down-tide, on a course parallel to the buoy but two or three boat-lengths off it. When the buoy is on the beam, a slow turn up into the wind and a final luff will bring the buoy under the bow. When the buoy is 'plucked'; the sails should be quickly dropped.

With wind over tide the approach is made against the tide, with the wind on the quarter or on a downwind run. Because you cannot spill wind from the main in these circumstances you must approach under jib alone. The strength of the tide is again crucial.

On some occasions the wind will blow across the tide: approach, if the tide will let you, under main alone on a broad reach, rounding up and easing the sheet to bring you neatly up to the buoy.

Mooring under sail is a skill worth acquiring, because as we have said, the engine can fail; however, be prepared to have several attempts, and always have an escape route. Furthermore prepare both the boat and the crew: a good exercise is to spend time explaining your plan of attack and how you want the crew to behave.

Mooring Alongside

Mooring alongside a large floating pontoon or a long harbour wall uncluttered by other boats, in calm waters and with

Early morning, lines of moored yachts.

little wind and no tide, is a relatively simple manoevre, the only real consideration being prop-walk. For a right-handed prop, go alongside port to; approach at some 20 degrees to the landing at slow speed. When almost alongside the helm will give a short but powerful burst astern, causing the boat to stop her forward motion and pivot her stern to port. The crew step onto the pontoon with breast ropes, and, when these are secure, they follow with springs and longer fore and aft shore warps if necessary.

Should it be decided to come in starboard side to, under the above conditions, then at the final approach the helm should turn the bow outwards so that the midship and quarter of the yacht are nearer to the pontoon than the bow, with the quarter only a few inches off it. Applying the short but powerful burst astern will again stop the forward motion

of the yacht, and the bow will pivot inwards towards the pontoon. In general, some wind and stream will be present as you approach, so the accepted practice is to approach from downstream – that is, into the tide – to gain advantage of its stopping effect. If in any doubt as to which is the stronger, wind or stream, stop the yacht some little way off the pontoon so as to experience the combined wind/tide effect upon you, do a dummy run, and take your time.

If your approach is correct, then your aiming point will remain dead ahead; if it is not, your yacht will 'crab' across her heading: crabbing to a point in front of your heading means that you are being 'carried' and are not able to take advantage of the tide/stream braking effect. If you are crabbing to a point behind your heading, then your speed through the water needs to be increased.

3

CHARTS: AN INTRODUCTION

There are many skills that the mariner must acquire, and among the most important is the ability to understand and use three basic aids: the chart, the magnetic compass and the log. Each of these will have a chapter devoted to it, starting with the chart.

A chart is a flat representation of some named part of the curved surface of the world. In this chapter we will cover only a small part of the immense amount of information contained on a chart, and will start by reminding you that any position on the earth's surface may be defined by its latitude and longitude.

Latitude

The latitude of a position on the earth's surface is measured in degrees and minutes, north or south of the equator. Technically it is the angular distance of that position north or south of the equator, measured from the centre of the earth. All places having the same latitude are said to lie on a parallel of latitude. Fig. 30 illustrates the 30°N and 60°N parallels of latitude.

The poles of the earth are 90°N and 90°S of the equator. Each degree has sixty minutes, and for extra accuracy, tenths of a minute may be used in specifying the latitude of a place on the earth's surface.

Longitude

The longitude of a place on the earth's surface is also measured in degrees, minutes and tenths of a minute. It is measured east or west from Greenwich, having a maximum value therefore of 180°(E or W).

A line joining the two poles and passing through a place is called the meridian of longitude of that place, or more simply its meridian. Fig.31 shows the meridians of 20°E and 30°W. The meridian of Greenwich is by definition 000°.

All meridians of longitude, if extended to encircle the world, would form a *great circle*, this being one whose plane bisects the earth. Compare meridians of longitude with parallels of latitude: the only parallel of latitude lying on a great circle is that of the equator, all the others are said to be small circles.

It was stated earlier that any position on the earth's surface can be defined by its latitude and longitude: for example, the position of the west entrance to Dover

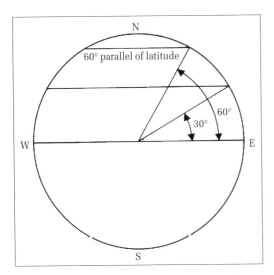

Fig 30 Illustrating the 30° and 60° angles from the centre of the earth to its surface. These angles define the two parallels of latitude 30° North and 60° North. All parallels of latitude, except the equator, are small circles, that is they do not bisect the globe of the earth.

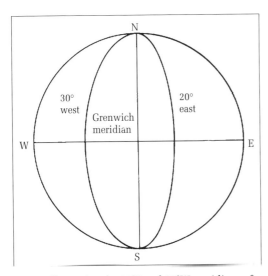

Fig 31 Illustrating the 20°E and 30°W meridians of longitude. Longitude is measured East and West of Greenwich. Meridians of longitude if extended to completely circle the earth would form 'great circles', that is circles which bisect the earth. The equator is the only great circle of latitude.

harbour is 51° 06'.7N 01° 19'.8E. This statement will hopefully be confirmed by you later in this book.

The Mercator Chart

There are many methods used to project the earth's curve onto the flat representional chart; initially we will be referring to the Mercator projection which, as shown at Fig. 32, encloses the earth in a vertical cylinder. Angles of latitude from the centre of the earth are extended beyond the earth's surface to strike the cylinder; meridians of longitude are like-

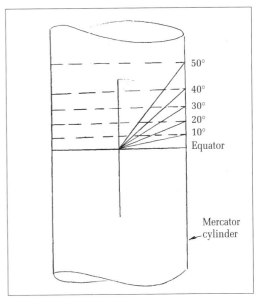

Fig 32 The 'Mercator cylinder' enclosing the globe of the earth, in the mercator projection the 'angles of latitude' are extended to the cylinder walls to form the Mercator parallels of latitude. Notice the non linearity of the scale, the charted distance between the parallel lines of latitude increases with increasing latitude.

Meridians of longitude are also extended to the cylinder, in doing so the lines of longitude are represented as equidistant parallel lines.

wise extended. Finally, all the earth's topographical features are also extended. When the cylinder is unrolled to become the flat Mercator chart, the earth's features are overlaid with a lattice of latitude and longitude lines.

You should note the inherent distortion in the Mercator projection, viz. parallels of latitude remain parallel, but become further apart as latitude increases. This distortion is not of paramount importance to the European sailor, but it becomes increasingly important at high latitudes and is one of the reasons why other methods of chart projection are used. A second distortion is due to the fact that meridians of longitude on the Mercator chart remain parallel at all latitudes.

One nautical mile – or sea mile, usually abbreviated 'n. mile' – is the equivalent of

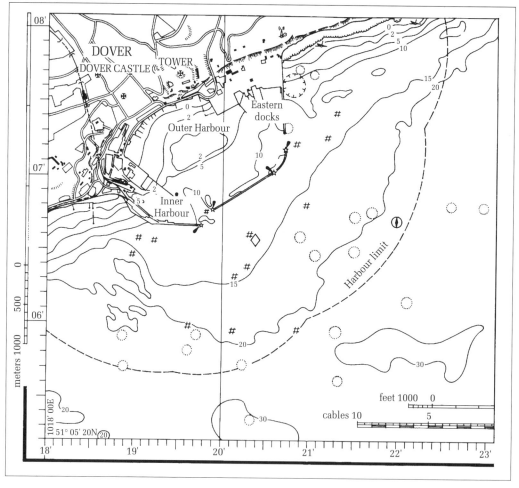

Fig 33 The bottom left hand corner of a chart; note the small numbers in the corner; 51°05.20N and 1°18.00E defining the position of that corner in latitude and longitude. The vertical latitude scale is annotated in one-minute intervals, sub-divided into 10ths of a nautical mile.

one minute of latitude measured along a meridian at latitude 48°. The nautical mile is thus approximately 6,080ft, slightly longer than the land mile. One tenth of a nautical mile is called a 'cable', and is approximately 608ft or 200yds. A ship's speed is measured in knots, where one knot equals one nautical mile per hour.

Distances on a Mercator chart are measured along the latitude scale, that is along the left-, or right-hand vertical scales of the chart (see Fig. 33); furthermore, due to the inherent distortion of the Mercator projection, distances should be measured on the latitude scale alongside your position and not on the scale at higher or lower latitudes.

When using Mercator charts covering distances of 600 nautical miles or more, the distortion in the latitude scale can be seen by comparing one minute of latitude at the lowest latitude to one minute of latitude at the highest latitude shown on the chart, you should notice a small but measurable difference. On the Mercator chart, the meridians of longitude appear as equidistant parallel lines lying north and south. The scales of longitude – appearing at the top and bottom of the chart and measured in degrees and minutes – represent linear distances on the earth's surface, and, because the meridians converge at the poles, these linear distances decrease with increasing latitude. In other words, one minute of longitude at a particular latitude represents a greater distance in nautical miles than does one minute of longitude at a higher latitude. It follows that distances on a Mercator chart are 'always, without exception and in all circumstances' measured along the latitude scale.

Chart Bearings and Courses

The direction of a yacht's course and all bearings plotted on a chart are measured in degrees true, ie from the geographic north pole in a clockwise direction, using three-figure notation:

True North = 000°
Due East   = 090°
Due South = 180°
Due West   = 270°

Figure 34 shows a compass rose with an outer circle in degrees true, and an inner circle marked off in degrees magnetic; more about the magnetic circle later. At least one rose appears on all charts.

One other method of specifying a direction is to follow the old but useful system of points. This system is based on the four cardinal points of the compass, viz. North, South, East and West; each quadrant then is subdivided into eight points, each of 11¼° and named; for example, in the quadrant north-to-east the points are:

N, NbyE, NNE, NEbyN, NE,
NEbyE, ENE, EbyN, E

A full picture of the point system is shown at Fig. 35. The system is used for wind direction in weather forecasting, and is therefore useful knowledge. To recite in succession the points of the compass is said to be 'boxing the compass'.

Chart Instruments

People have been known to work on a chart using a pencil, a simple ruler and a protractor. However, to make life easier the following instruments are essential:
1. A soft 2B pencil and a soft rubber. Charts are expensive and need treating carefully: every line you plot will eventu-

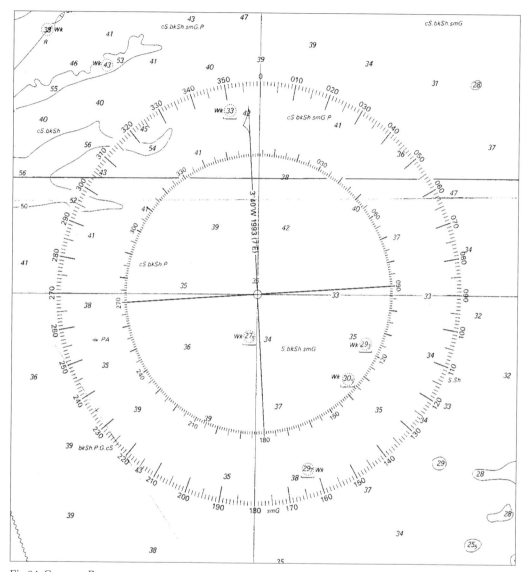

Fig 34 Compass Rose.

ally have to be (gently) erased, so lines should be only as long as is necessary, and clearly but not heavily drawn.

2. A pair of brass dividers; these are specially made for distance measurement when plotting on the chart, and can be used with one hand.

3. One of the following: parallel ruler, Breton plotter, rolling ruler, Douglas protractor or Hurst plotter. Unfortunately there is no 'best instrument', and users have their favourite instrument; mine is a pair of parallel rules with angle gradations around its periphery, although I must confess that

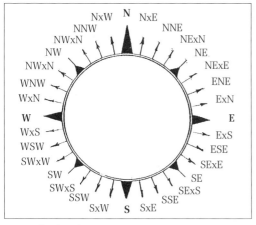

Fig 35 Boxing the Compass.

Sailing data and instruments.

occasionally I have cursed when slippage or 'something' under the chart has obliged me to start again. A number of these instruments are shown above.

Chart Plotting

Fig. 36 shows a charted area around Dover; it is copied from Admiralty Chart no. 5061, entitled Dover to North Foreland. You should now be able to confirm that the western entrance to this port is indeed 51° 06'.7N 01° 19'.8E. Using whatever implements are available to you, determine, using Fig. 36, the latitude and longitude of the eastern entrance to Dover.

A second method of fixing a position on a chart is to take a bearing and distance from a known conspicuous shore object. Fig. 36 shows such a bearing line taken from the western end of Dover's southern breakwater to a point X: the bearing is measured using the compass rose on the chart, and the distance is measured by taking the length of the line to the adjacent latitude scale using a pair of dividers. The line bears 130° from the eastern end of the southern breakwater and the distance to point X is 0.8 n. miles.

Charted Depths

A chart has many useful, if not essential, items of information to aid the safe passage of sea-going vessels. For example the depth of water may be shown from shoreline to deep water; Fig. 37, again taken from chart 5061, shows a sea area near the Goodwin Sands with many numbers printed on it, and these values indicate the depth of water at that position for the lowest predicted tide.

The area called 'South Calliper' of Fig. 37, coloured green on the original chart, has a special significance in that it is dry at the lowest possible tide. Its periphery is marked by a change of colour and a broken line with the digit 0 embedded in the line; the latter means that the periphery of South Calliper would be awash at the time of the lowest predicted tide. The numbers shown on South Calliper itself are of the form $X\underline{y}$, and should be read as X.y metres; they predict how high above the

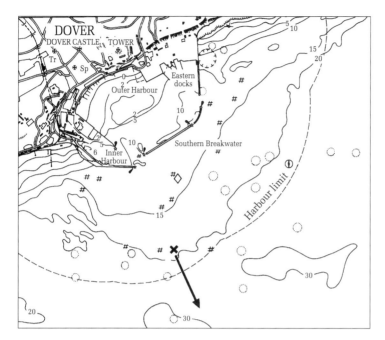

Fig 36 Chart details of sea area Dover. Point X is a bearing line, 130°T, 0.8 n. miles from the seaward end of the western breakwater. Note the 15 and 20 metre depth contours outside the breakwaters and the 2, 5 and 10 metre depths inside the harbour walls. Note also the conspicuous land features shown; Dover castle, Sp(ire) and the tower with its height shown alongside (131 metres).

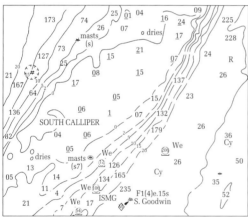

Fig 37 Sea area South Calliper, showing drying heights and surrounding depth contours.

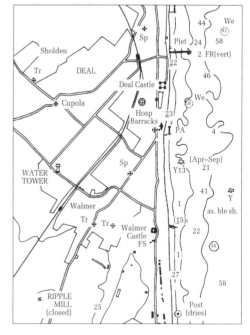

Fig 38 Coastline of approaches to Deal. The depth contours off the coast indicate a gently sloping seabed.

water that point will be at the lowest tide. For example, the value 1_7, shown on South Calliper means a height of dry land 1.7m above the surrounding waters at the lowest predicted tide. It follows that the immediate area off South Calliper is covered with water at all states of the tide.

Depth Contours

A series of continuous lines called contour lines can be seen surrounding the drying land of South Calliper; each line is marked with a single digit – 2, 5, 10, 15 or 20 – and this indicates the depth in metres of that line. Contour lines join together depths of the same value; further out is a 30m line.

Contour lines can be of great value to a fog-bound mariner feeling his way along a safe but unseen coastline. This last statement is not intended to encourage sailing in foggy conditions, but to indicate to you that, if you are ever caught out in fog, contour lines off a gently sloping shoreline may be of navigational assistance. Fig. 38, copied from Chart 5061, shows the land-

mass, drying heights and charted depths of the approaches to Deal.

Shore Features

Conspicuous land features such as coast-guard stations, water towers, churches, hotels and lighthouses, from which it may be useful to take bearings, are also shown on the chart. A word of warning: you must be absolutely sure that the shore feature from which you are about to take the bearing is the one you are going to plot from on the chart – there may be several churches in the area of interest, and if you are in doubt, choose another feature for your bearing.

Chart 5011

Chart 5011 is actually a booklet, and it illustrates and has examples of every special feature that may appear on your chart. Fig. 39 shows just a few examples of how these special features are drawn, and explains their significance.

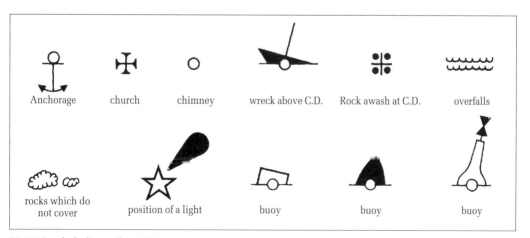

Fig 39 Symbols from Chart 5011.

Buoyage

In 1514 Trinity House was given a Royal Charter to regulate the laying of buoys around England. The first buoys were laid in the estuary of the River Thames, and consisted of baulks of timber and wooden barrels. Three centuries later a lateral system of buoyage was introduced by Trinity House, in 1882; this determined that can-shaped buoys would be passed on the port-hand, conical-shaped buoys on the starboard-hand, and that spherical buoys would mark middle ground. This system was adopted around Britain and later extended to the British Empire. Then in 1889 an international conference in America recommended a second lateral system in which colour rather than shape was the important feature, starboard-hand buoys to be painted red, port-hand painted black or parti-coloured.

In 1870 some thirty different systems of buoyage existed around the world and confusion and mistakes were the danger-ous results of these multi-standards. Fol-lowing a series of accidents in the Dover Straits in which fifty-one lives were lost, a convention in 1973 of the International Association of Lighthouse Authorities (IALA) recommended an international system of standardized buoyage. A simple unambiguous system, the IALA Maritime Buoyage System 'A' was adopted by the countries of north-west Europe. Only five types of buoys, or marks, are defined, and they may be used in any combination:

Lateral marks: Indicate the port and starboard limits of defined channels.
Cardinal marks: Indicate which cardinal point of the compass relative to a danger is the safe direction to pass.
Safe-water marks: Indicate safe water at, and all round the mark.
Isolated danger marks: Indicate an isolated danger at the mark, with safe water all round.
Special marks: Indicate an area of note or feature; reference to the chart will reveal a special mark's significance.

Lateral Marks

The basic principle of lateral marks in System A is that port- and starboard-hand marks are linked to the 'general direction of buoyage' which usually flows clock-wise around continental land masses. Within river estuaries, however, buoyage direction is from seaward.

Examination of the relevant section of the 'in use' chart or pilot will allow the prudent mariner to list the buoyage mark-ing his inshore passage. Where doubt exists, the chart will carry a large magenta arrow indicating the general direction of buoyage. You will also need to be aware that in some places the general direction of buoyage may reverse, for example at Egypt Point (IOW). Where the general direction is reversed, vessels may approach this area from the Needles, that is, from westward; or from the Nab Tower, that is, from eastward, moving along equally important and main channels (see Fig. 40).

Channel marker.

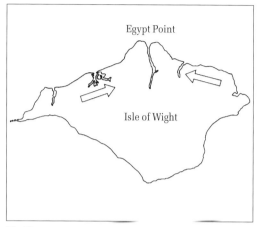

Egypt Point

Isle of Wight

Fig40.

It is no bad thing to have available in the cockpit a list of the buoys, between berth and open sea, with the distances and courses between them. Fog or banks of fog can come up quite quickly, and the list, together with echo soundings, can prove invaluable.

In system 'A', port-hand marks are solid, red, can-shaped buoys or vertical painted poles: the cans may, and the poles must carry a red, can-shaped topmark. Starboard markers are solid, green,

conical-shaped buoys or vertical poles, the cones may, and the poles must, carry a green conical topmark. Fig. 41 illustrates and lists the typical shapes, topmarks and colours of lateral marks in system 'A'.

Cardinal Marks

Cardinal marks are named after the side of the danger or feature at which they are placed; the name therefore indicates the safe side on which to pass. The colouring is yellow and black banding, and a topmark is always fitted. The buoys are named north, south, east and west cardinals; Fig. 42 shows how they may be positioned around a dangerous shoal area. Four marks are used:

North cardinal marker (NCM):

pass north of it.

South cardinal marker (SCM):

pass south of it.

East cardinal marker (ECM:

pass east of it.

West cardinal marker (WCM):

pass west of it.

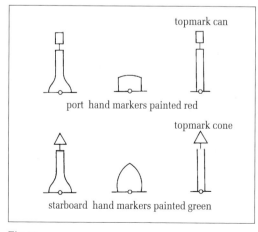

topmark can

port hand markers painted red

topmark cone

starboard hand markers painted green

Fig 41.

North Cardinal marker.

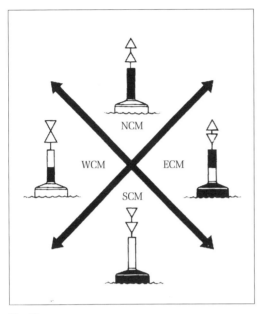

Fig 42.

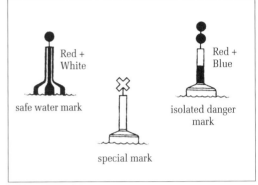

Fig 43.

Safe Water Mark

Fig. 43 shows this most welcome of all buoys; it is often used as a landfall buoy, indicating to the homeward-bound sailor the safe seaward end of a channel. Colours are vertical bands of red and white, and the buoy always carries a single red ball topmark.

Isolated Danger Mark

Marking a single danger with safe water all round, the colours of red and black banding give this buoy a most forbidding appearance. It always carries a black topmark of two vertically displaced black balls.

Special Mark

These buoys are not specifically for safe passage indication, but are often used for military exercise areas or to mark spoil grounds. They may be of any shape, and

Leading marks.

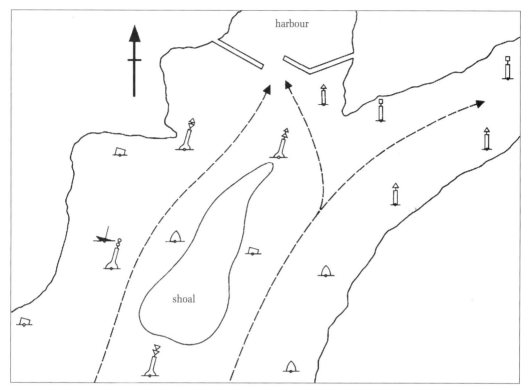

Fig 44.

are always painted yellow with a cross, as is the compulsorily fitted topmark.

Lights

The very first thing to say about lighting is that it isn't always fitted to navigational marks. Many buoys are unlit; they are not necessarily 'yacht-attracting', as if by special plastic magnetism, but they do often seem to pass close to a yacht at night time. When planning an inshore night passage, such buoys need to be specially noticed and a watch kept in their vicinity.

Lateral marks are lit by the appropriate red or green colour; the pattern of light characters can be of any patterns. On cardinal marks, lights are white when fitted:
NCM, continuous white flashing.
ECM, three short flashes.
SCM, six short flashes plus one long flash.
WCM, nine short flashes.
On safe-water marks, the white light may be isophase, of equal on/off periods; occulting, where the 'on' periods are shorter that the 'off' periods; one long flash every ten seconds; or morse letter 'a'.

Fig. 44 is one example of how a combination of buoys may be used in a channel.

4

THE MAGNETIC COMPASS

The magnetic compass is the means by which the yachtsman controls the direction of his boat across the seaways and oceans of the world. It is used in conjunction with the chart and log to ensure that a calculated, known and safe passage is followed from the yacht's departure point to its destination.

It is the helmsman's job to steer the compass course given to him by the navigator, and to make this task even

Ship's bulkhead-mounted compass.

remotely possible the course given is rounded to the nearest 5°. On long passages the helm should be changed at, say, two-hour intervals, because the required concentration is difficult to maintain, especially at night when the compass is lit at only a low level. Once the yacht has settled onto her course, the helm will find this easier to maintain if, instead of concentrating solely on the compass reading, a 'look ahead' is taken of a distant shore feature (where possible), referring to the compass just occasionally to check his course.

On a chart, the meridians of longitude – theoretically converging on remote geographical North and South poles, whose true bearings are respectively 000° and 180° – are shown parallel and equidistant. The outer circle of the compass rose on the chart has the reference bearing of 000°(T), in other words the compass rose is aligned to the remote geographical North pole, and all true bearings are referenced to this alignment. Unfortunately for the yachtsman, the needle of a magnetic compass points to a magnetic north pole, and this is not in the same position as the true pole. The angular difference between the true and magnetic pole is called 'magnetic variation'.

Magnetic Variation

Measured from the UK the magnetic pole is offset from the true pole by some 3° and slowly, over many years, rotates around the true pole (see Fig. 45). In UK waters the earth's magnetic north pole is 'seen' to be a few degrees to the west of the earth's geographic or true pole, which means that a magnetic compass will read slightly higher than the true bearing. This error is called magnetic variation, and in UK waters is approximately 3°W, meaning that the magnetic pole is 3° west of the true pole. A yacht steering 000° by a magnetic compass is therefore heading 3° west of true north.

A study of Fig. 46 will show that the value of magnetic variation is position dependent – it is zero at two particular opposite meridians – whilst at all other positions on the earth's surface magnetic variation will be a few degrees east or west. In practice, therefore, and as we shall see later, when the navigator draws a line on the chart joining his departure and destination points, he is able to read his course from either the true compass rose (XXX°T) or the magnetic compass rose (YYY°T).

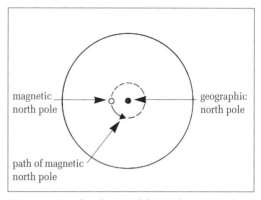

Figure 45 Overhead view of the earth's Geographic North Pole (GNP) with the earth's Magnetic North Pole shown off to one side. The MNP rotates very slowly around the GNP at some seven minutes of arc each year.

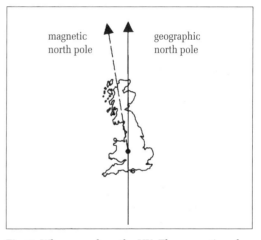

Fig 46 When seen from the UK. The magnetic pole is approximately 3°W of the true pole. When seen from the opposite side of the world (longitude 180°) the magnetic pole would be some 3°E of the true pole.

Variation, due entirely to the misalignment of the geographic and magnetic poles, is the reason for the inner 'magnetic circle' of a chart's compass rose being offset, as shown at Fig. 34 (p48); note also on Fig.34 that the value of variation for a specific year is shown together

Hand bearing compass.

with the annual change in that variation. In Fig. 34 the figures 3° 40'W 1993 (7'E), mean that in this position the magnetic pole was 3° 40' west of the true pole in 1993, and moving 7' easterly each following year, ie in this case decreasing 7' annually at that time. It follows from this that a magnetic bearing of a shore object taken in 1993 would be 3° 40' bigger than the true bearing of that same object. A course of 046° true could also be quoted as 049° 40' magnetic, and would be rounded up to 050°.

Magnetic Deviation

Magnetic deviation is a second compass error: it is the effect upon a yacht's compass of local magnetic effects, and the presence of ferrous materials on board the yacht itself; the inboard engine, electrical circuitry and iron ballast keels are just some of the items that will permanently pull the magnetic compass away from the magnetic pole. The temporary positioning of tools and/or metal cans in the vicinity of the compass will also have an effect, pulling the compass one way or the other.

A moment's thought will show that deviation is further complicated in that it will change with the yacht's heading, that is, it may well pull the compass off to the east on one heading, to the west on a second, yet have no effect on a third. Fortunately the deviation for each yacht can be measured, reduced to a minimum, and a chart drawn showing the value of deviation for each heading.

The Deviation Curve

A typical deviation card is shown at Fig. 47. The measured value of deviation is

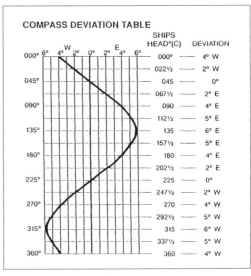

Fig 47 Deviation Card.

shown for each 45° of ship's heading. Notice that the curve swings east and west of zero; for example, on a ship's heading of 090° the deviation is approximately 4°E, indicating that the compass is being pulled 4° east of the magnetic pole; whilst a heading of 315° produces 6° west deviation.

You might assume that a yacht's steering compass has been swung by a professional compass adjuster, and a deviation card produced from that swing; and if this has *not* been done, you should ask yourself why not? Once done, the skipper should check the deviation of the yacht's main compass periodically; there are two relatively simple means of doing this:

1. By using the hand-bearing compass to check the course being steered by the main compass. Stand in the forepart of the yacht to do this, as far as possible away from any magnetic effects.

2. By heading along a charted transit where the transit bearing is given on the chart, or make up your own transit line(s)

from conspicuous shore objects and check your course as you steer along it.

Total Compass Error

The total compass error is the algebraic sum of variation and deviation. If both errors are named east, or if both are named west, simply add them together and name the resultant total error east or west respectively. If the errors are named differently, then subtract the smaller from the larger and name the result after the larger of the two.

Example 1
If variation is 3°W (the magnetic pole is 3° west of the true pole) and deviation is 4°W (the compass is being pulled 4° west of the magnetic pole) the total error is 7°W. This means that a compass bearing or course would read 7° bigger than the true bearing or course.

Example 2
If, at a particular place and on a specific course, variation is 6°W and deviation 3°E, the total compass error would be (6°–3°)W = 3°W and the compass course would read 3° higher than the true course.

We are now in a position to define a yacht's course in one of three ways:
1. A course in degrees true: taken from the true pole, it is abbreviated (°T).
2. A course in degrees magnetic: ie allowing for variation, taken from the magnetic pole, it is abbreviated (°M).
3. A course measured in degrees compass: ie allowing for variation and deviation, taken from the compass pole, it is abbreviated (°C).

When a course line is drawn on a chart and annotated, the convention is to use the true figure; this means that the (°T) may be omitted. In consequence, if a navigator insists (for some strange reason) on annotating the course line using the value obtained using the definitions of either 2 or 3 above, then it is essential that the appropriate (°M) or (°C) is used.

Please complete the following exercises before leaving this section:
Q.1. Assuming that variation is 4°W, calculate the true course given the following data:
    a. Compass course 025(°C)
                              deviation 2°W
    b. Compass course 140(°C)
                              deviation 5°E
    c. Compass course 260(°C)
                              deviation 4°E
Fig. 48a,b and c show the three pictorial representations of questions 1a to 1c above; however, please do not study them until you have attempted Q.1 above.
   In answering Q.2 below, it may help you to draw similar diagrams in the space below before calculating the answers to Questions 2a to 2c.
Q.2: Calculate the compass course to steer given the following data:
    a. True course   030(°T) deviation 3°W
    b. True course   160(°T) deviation 6°E
    c. True course   270(°T) deviation 4°E
Space for your answer to Q.2 above

Finally on this subject, two little odes:

1. **Total error west, compass best**.

ie If total compass error is west, the compass reading will be bigger than the true one.

2. **Total error east, compass least**.

ie If total compass error is east, the compass reading will be smaller than the true one.

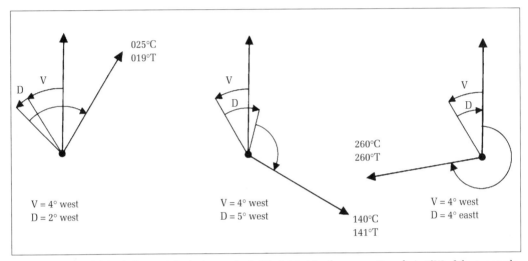

025°C
019°T

V = 4° west
D = 2° west

260°C
260°T

V = 4° west
D = 5° west

140°C
141°T

V = 4° west
D = 4° eastt

Fig 48 Diagram (a) Variation and deviation are both West. That is, the magnetic pole is 4°W of the true pole and the compass is 2°W of the magnetic pole. Total error is therefore 6°W. The true course is 025–6=019°T.

Diagram (b) Total error is 5°E–4°W=1°E.
     The true course is 140+1=141°T

Diagram (c) Total error is zero.
     The true course is 260°T.

5

CHARTWORK 1

We have discussed two of the mariner's three basic aids, the chart and the magnetic compass, and have established that the chart provides all the necessary information about our surroundings, the depth of water, navigational hazards and safe transit lines being only a small part of a chart's data. Basically it allows the navigator to plot a safe ground track, or route, from departure to destination. The magnetic compass, its course corrected for variation and deviation, allows the helm to steer along the navigator's charted true course, thus ensuring a safe passage. Our third basic aid is the ship's log, and this instrument enables the mariner to calculate the distance run and hence the speed of his yacht through the water.

The Log

Originally the log consisted of a paddlewheel protruding through the hull, the inboard end of which contained a box full of pebbles; as the paddle turned, pebbles fell into a container and were counted. Nowadays it is usually either an analogue taffrail 'Walker log', or an electronic one.

The Walker log comprises a mechanical counter with analogue dial mounted on a special bracket at the stern of the boat. The counter is driven by a propeller trailed behind the yacht at the end of a long cord; when the boat moves through the water, the propeller spins and the cord transmits the spin to the counter. It is worth mentioning that the recovery of the Walker's propeller after use needs to be done in a particular way. When pulling in the long trailing cord, allow the recovered part of the cord to fall back into the water until the propeller is on board, and only then retrieve the cord completely; in this way the rotating propeller will not snare up the cord during the recovery operation. If you forget this advice it will come back to you in a flash when you attempt your first Walker log retrieval.

An electronic log consists of an under-hull transducer, which may be entirely electromagnetic or have an external mechanical impeller, together with a digital dial fitted at the navigator's position in the cabin and probably a repeater dial in the cockpit. The electronic log is most convenient since all we have to do is switch it on and read the dial. However, if it is of the type with an under-hull-fitted impeller, the one piece of essential maintenance required is to clean the impeller from time to time. In order to do this the transducer is removed from its hull-piercing fitting in the bilge, and a substitute bung is immediately and firmly placed into the transducer fitting to halt the inrush of water. Once cleaned of any marine debris, the process is reversed and the transducer refitted.

Before a voyage begins, the prudent skipper will have studied the appropriate material, gathering information on such things as the likely wind strength and direction, times of high and low water, the predicted weather and many other details. Once under way, pilotage down the river will take the boat to open water, and now for the first time the skipper's predictions of the wind, sea state and visibility can be judged. Hopefully his good expectations are confirmed, and so he can plot his position, hoist his sails and set his first course.

Plotting a Position

Good chartwork, accurately and neatly plotted is the hallmark of good seamanship, and a mariner must strive con-

stantly to calculate and then plot his position. The latitude and longitude of his departure point, the time of departure and log reading, all this should be noted, charted and annotated.

When leaving a harbour or a well marked channel the departure point is easily recognized on the chart and a 'fix' plotted, as for example '2 cables due south of the harbour entrance', or 'the fairway buoy 2 cables to starboard'; these are quite adequate chart fixes (see Figs. 49 and 50). A charted fix is shown as a dot inside a small circle, and it should be annotated with the time and log reading. Don't forget to enter these details in the yacht's narrative log.

If your point of departure is a small bay, then two or preferably three conspicuous charted shore objects should be found, and bearings taken of them using the hand-bearing compass, as shown in Fig. 51. Each bearing when plotted gives a position line, so called because your yacht's position must be somewhere on each of these lines. It follows that the three plotted

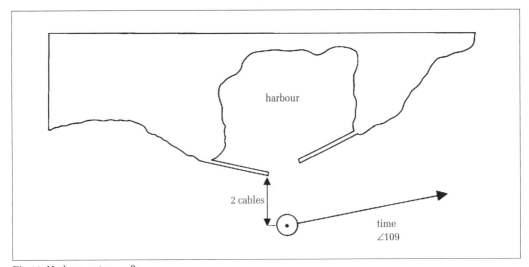

Fig 49 Harbour entrance fix.

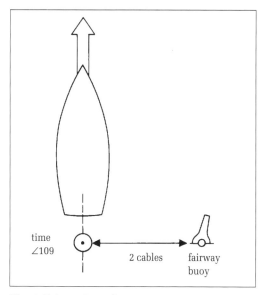

time
∠109

2 cables   fairway
              buoy

Fig 50 Fairway buoy fix.

position lines must cross at one point, and this point fixes the position of your yacht at the time the bearings were taken.

In practice the three position lines will rarely cross at a single point but are most likely to cross over a small triangular area called a 'cocked hat' (see Fig. 52). This positional discrepancy usually occurs because of the difficulty in taking bearings from a platform which is moving simultaneously in all three dimensions.

Plotting a Course

Assuming that your voyage is reasonably short and that both departure and destination ports appear on the same chart, your next operation is to pencil in lightly

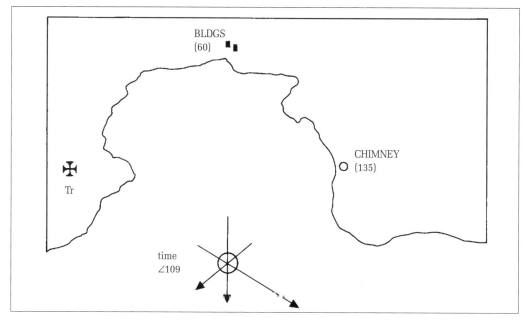

Figure 51 Three-position fix. Shore objects are conspicuous:

1. Tower
2. Buildings, height 60 metres
3. Chimney 135 metres.

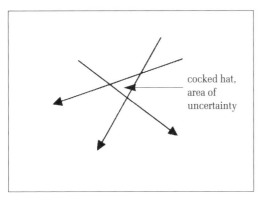

Fig 52 Cocked hat of three-position fix.

a line between the two ports. This line is known as the 'ground track', and is the exact course which you intend to follow in order to get to your destination.

Two questions must now be answered:
1. Is the ground track a safe one to follow, ie a distance of at least half a mile separates it from all dangers, such as shallow water, wrecks and rocks?
2. Will the wind direction allow you to sail it?

Assuming 'yes' to both these questions, the true bearing of the ground track is then measured and converted into degrees compass in order to give the helm a course to steer. In case you have forgotten how to do this, here is a little exercise for you to do: in preparation for a sailing passage, a navigator noted the true bearing of a plotted ground track, measured on the chart's compass rose as 165°(T). Given a magnetic variation of 3°W and magnetic deviation obtained from the card at Fig. 47 (p.58), what compass course was given to the helm? (The answer is given later, but readers please calculate it now!)

With a known compass course, details can be entered into the narrative log, the sails trimmed, and the journey can begin.

Dead Reckoning Position

Known as the DR position, this is a plot on the chart which takes into account the compass course steered and the log distance run during the previous hour. A DR position is the least amount of information that must be noted and charted at each hour of a sea passage.

In the above example the compass course was 164°, and if we assume that the log at the start was 345 n. miles and after the first hour was 351 n. miles, the distance travelled from our departure point through the water was 351–345 = 6 n. miles. Thus the first DR position of the journey can now be put on the chart, and is conventionally plotted by means of a small mark cutting the 'course steered' line. Alongside the DR mark is written the time and log reading. (A DR position is shown at Fig. 53; please note that to avoid confusion between the time given using the twenty-four hour clock, and the log reading, the log reading is written inside a long-bottomed L.)

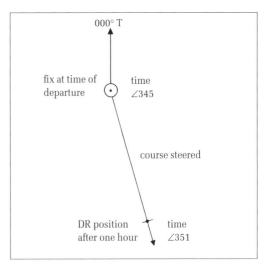

Fig 53 Dead reckoning DR position.

6

TIDES

The tidal rise and fall of the sea level around our shores is the result of the gravitational effects of the moon, and to a lesser extent the sun. The moon's greater effect is because it is very much closer to the earth than the sun, even though it is much smaller.

Lunar and Solar Tides

The moon is the biggest single factor controlling the times and heights of the tides. Considering the moon's influence, as the earth spins on its axis once every twenty-four hours, the sea and ocean waters are quite literally lifted towards the moon by its gravitational pull. The result is a rise in the level of the water immediately below the moon, and in the waters immediately opposite, on the other side of the world. The sea waters in areas at 90° to the moon's position will fall in height in order to supply these rises. Fig. 54 shows the resultant bulges in the sea immediately below the moon and on the opposite side of the world.

The moon orbits the earth every 27.3 days. The world, in completing one of its own revolutions in 24 hours, must there-

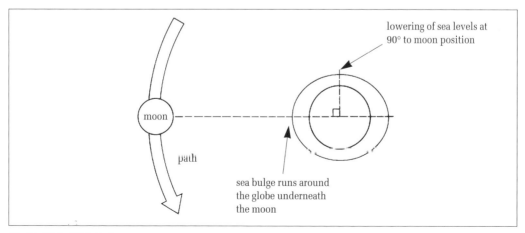

Fig 54 Global tide bulge.

fore turn a little further each day for the high-water point to occur at the same spot on earth. This means that the time of high water slips back a little each day. For each earth spin of 24 hours, the moon will have moved a further ½₇ of its orbit, and so the delay in each day's tides will be: (24/27)×60 = 52 minutes. This equation is only approximate, and if we take account of the fact that the earth is also moving along its orbit around the sun, the diurnal delay in the times of high water becomes nearer to 50 minutes.

In theory, during every 24-hour and 50-minute period there will be two high and two low waters, each high water occurring 12 hours and 25 minutes after the previous one. However, oceanic tidal waves are subjected to friction against the irregularly shaped ocean floor, resistance from continental shelves, and a great variety of underwater interferences and obstacles, and consequently the earth witnesses a wide variety of tidal behaviour.

The UK and European coastlines experience 'semi-diurnal' tides, the theoretically predicted two high and two low waters in each 24-hour 50-minute period. Other parts of the globe have only one high and one low water each day; some coasts have 12m (40ft) tides while others, like the Mediterranean, have only a few centimetres' variation.

The sun also affects the earth's tides; however, it is 93,000,000 miles (150,600,000km) away, and, although of immense size, its tidal effects serve only to modify the lunar tides.

The moon in orbit around the earth makes one complete revolution every 27.3 days; however, if the moon were new at the beginning of an orbit period (position (a) in Fig. 55), it would not be new at the end of that period (position (b))

because it would not be in a straight line with the earth and the sun. To achieve that position takes a further 2.16 solar days (position (c)) before the moon is new again. This means that a lunation – new moon to new moon, or a lunar month – takes approximately 29.5 solar days. From this it can be assessed that the full moon, occurring at the mid-point of a lunation, will be some 14.7 days after a new moon.

The half lunation period of 14.7 days is of great importance to the mariner since it represents the interval at which the sun and moon are in line with the earth. The combined gravitational effects of sun and moon are at a maximum at new and full moons, giving rise to the highest high tides and lowest low tides; such tides are called 'spring tides' and they occur at approximately fortnightly periods. Midway between new and full moon times, the sun and moon are in quadrature in the sky, that is, they are separated by 90° in the heavens. The sun and moon each now pull the earth's waters in different directions, resulting in tides having the lowest high water and the highest low waters – that is, the 'neap tides'. In UK waters, the height of HW (LW) following a neap tide gradually increase ,(decreases) to a maximum (minimum) some seven days later at a spring tide.

Spring Tides

These occur twice each lunar month when the earth, moon and sun are all in line: in this circumstance the combined gravitational pull of the sun and the moon produces the biggest change in sea level – that is, the highest high level and the lowest low level (position (a) in Fig. 56. The tidal range, that is the difference

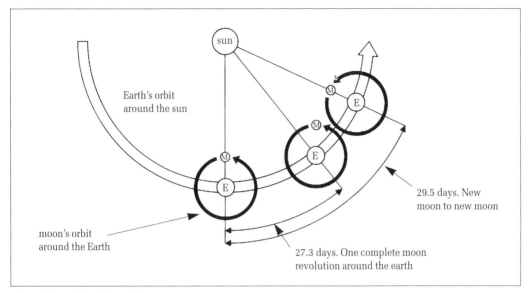

Fig 55 Lunar Month 29.5 days.

between the high-water (HW) level and the low-water (LW) level is therefore at a maximum. (Tidal range = HW–LW.)

Neap Tides

Neap tides also occur twice each lunar month, when the sun and moon are at 90° to each other (position (b) in Fig. 56); their gravitational effects are therefore acting at 90°, and the range of the tide (HW–LW) at this time is at a minimum.

Equinoxial and Solstice Tides

The earth's orbit around the sun is elliptical (see Fig. 57). It is furthest away at the solstices which occur on the 21 June and 22 December, and the sun's gravitational pull at this time will be at a minimum;

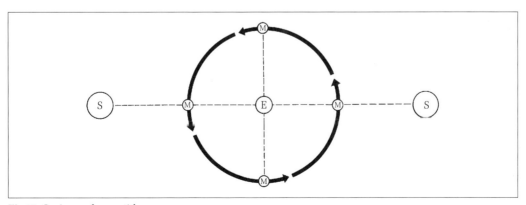

Fig 56 Spring and neap tides.

consequently the *solstice* spring tides will have a smaller range than the average spring tides.

The sun's minimum distance from the earth occurs at the equinoxes, that is, the 21 March and 23 September: the combined gravitational pull of the sun and moon will have maximum effect at this time, causing the equinoxial spring tidal range to be greater than the average spring tide range. The changing declinations of the sun and moon also have a spring tidal effect, such that very large spring tides may occur at equinoxes.

The importance of the high tidal range at an equinoxial spring tide is not that the HW height is at a maximum, but that the LW level is at a minimum. At times of exceptionally big equinoxial spring tides the LW height may be very low indeed,

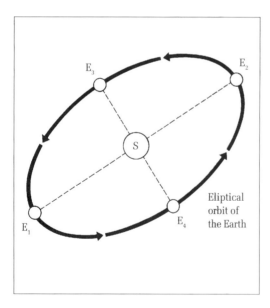

Fig 57 Equinoxial and solstice tides

E1 and E2 positions furthest from the sun give rise to the lowest spring tide ranges

E3 and E4 positions nearest to the Sun give rise to highest spring tide ranges

and under these circumstances the yachtsman returning to his normal berth, sailing unconcernedly in familiar waters, may find himself having to pause awhile until the tide turns.

Tidal Height Definitions

There are many names used to define the height of the tide (Fig. 63 shows some of them). Thus, 'ML' is the mean level, the average depth of water. 'MHWS' is mean high-water springs, meaning the average highest level of water average because, from our previous discussion of earth, moon and sun alignments, it follows that some individual spring HW levels would be higher or lower than the mean; for a similar reason, all the defined tidal heights shown in Fig. 58 define a band of heights about a mean figure. 'MLWS' is mean low-water springs, and defines the band of most concern to the mariner because it is the lowest level of water that he is likely to experience, and therefore the time when the greatest care must be exercised.

'MHWN' is mean high-water neaps, meaning the band of depths representing the lowest of the high-water periods, occurring when the sun and moon are in quadrature. 'MLWN', or mean low-water neaps, identifies the band of highest low-water periods.

'LAT', or lowest astronomical tide, sometimes referred to as CD, or chart datum, is the lowest predicted level of water and would only be reached at the low-water period of the most exceptional of spring tides.

We shall see later that the predicted times and heights of high and low water, for many ports, are given in published

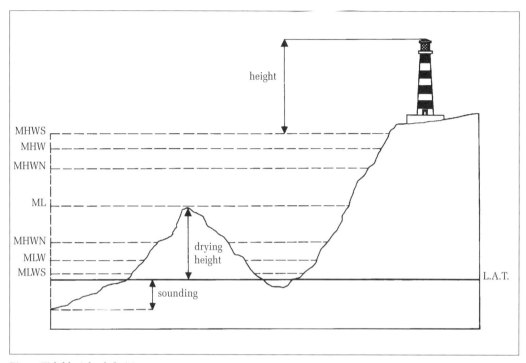

Fig 58 Tidal height definitions.

tide tables and almanacs. All tidal heights are measured from LAT.

Rise and Height of Tide; Depth of Water

These three expressions are often used during tidal height calculations, and refer to precise height datums, it is most important not to confuse them (see Fig. 59).

(1) The rise of tide, at any time, is the height of the sea surface above the previous low water. It is the height of water above the previous LW.

(2) The height of tide, at any time, is the sum of LW and the rise of tide. It is the height of water above LAT.

(3) The depth of water, at any time, is the sum of LAT and the height of tide. It is the height of water above the sea bed.

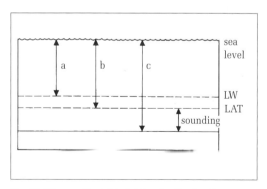

Fig 59 Rise and height of tide, depth of water

(a) is rise of tide
(b) is height of tide
(c) is depth of water.

Soundings

The depth of water below LAT is called the sounding and is the least depth of water predicted for that point. Soundings are those figures found in the water areas of all charts and are most useful. A mariner about to take passage, during good weather and on a calm sea, in a yacht drawing 2m, may do so safely in an area where the least sounding is 2.5m without further tide calculations. The extra 0.5m given here is to allow a safety depth for such things as wave troughs.

Fig. 58 shows an area where the sea bed rises above LAT, but which would not normally be exposed to view by tides at MHWS or lower. Such areas are called drying heights, that is, at some states of some tides the area will show above sea level. The height of such areas above LAT is printed on the chart in the form 3_7, meaning the height of this area is 3.7m above LAT.

'Drying heights' describe areas of land which will be exposed at some state of the tide. Land beaches are obvious examples, as the sea shoals towards the land, but drying heights may also rise above LAT as an isolated sea-bed hill, or rock; such areas require great attention from the mariner because they are areas of the sea bed which at some state of the tide, in clear weather, would be visible to him, but at other states of the tide may be anything from just awash to covered by a depth of water perhaps just less than his draught.

Heights of overhead cables and lighthouses are also shown on the chart; such measurements are taken with reference to MHWS.

Tide Tables

The semi-diurnal tidal heights and times for ports of the UK and associated waters are to be found in current copies of nautical almanacs. For tidal purposes, ports are defined as either 'standard' or 'secondary'. A standard port is normally large, has much commercial sea traffic, with many daily big ship movements. The almanac lists the predicted time and height of high and low water of all standard ports; examples are Dover and Southampton. We will define secondary ports later. Fig. 60 is a copy of the tidal predictions for the port of Dover during the months of May to August. The day, time and height of high and low water are listed for each day; for example:

Thursday 11 June

Time	m
0059	1.7
0617	5.4
1328	2.0
1836	5.6

From this you will see high waters occur at 0617 and 1836 Greenwich mean time (GMT), and that the heights are 5.4 and 5.6m above LAT respectively. Low waters occur at 0059 and 1328, and the heights are 1.7m and 2.0m above LAT respectively.

The range of the morning tide (HW–LW) is 5.4–1.7=3.7m; the range of the afternoon tide is 5.6–2.0=3.6m.

Please note that these height figures are measured above LAT, and a skipper in a yacht drawing more than 1.2m would need to study the charted soundings before entering at low water on the morning tide. Where these soundings plus LW are too shallow, the skipper would have to delay his port entry to a time at which the rising tide provided sufficient depth.

ENGLAND, SOUTH COAST — DOVER

Lat 51°07' N Long 1°19' E

TIMES AND HEIGHTS OF HIGH AND LOW WATERS

TIME ZONE UT (GMT)
Summer Time add ONE hour in non-shaded area

MAY

Day	Time	m		Day	Time	m
1 F	0327 / 0826 / 1553 / 2049	1.2 / 6.0 / 1.2 / 6.3		16 Sa	0447 / 0939 / 1706 / 2149	1.2 / 5.9 / 1.3 / 6.2
2 Sa	0424 / 0912 / 1647 / 2135	0.8 / 6.4 / 0.9 / 6.6		17 Su	0525 / 1013 / 1740 / 2226	1.1 / 6.1 / 1.2 / 6.3
3 Su	0516 / 0957 / 1736 / 2220	0.6 / 6.6 / 0.7 / 6.8		18 M	0556 / 1048 / 1807 / 2302	1.1 / 6.2 / 1.1 / 6.4
4 M ●	0604 / 1044 / 1821 / 2306	0.5 / 6.8 / 0.6 / 6.9		19 Tu ○	0624 / 1122 / 1838 / 2336	1.0 / 6.3 / 1.0 / 6.4
5 Tu	0649 / 1133 / 1904 / 2354	0.4 / 6.8 / 0.5 / 6.9		20 W	0656 / 1154 / 1912	0.9 / 6.3 / 1.0
6 W	0733 / 1224 / 1949	0.4 / 6.8 / 0.5		21 Th	0005 / 0730 / 1224 / 1947	6.3 / 0.9 / 6.3 / 1.0
7 Th	0043 / 0818 / 1314 / 2034	6.8 / 0.5 / 6.6 / 0.6		22 F	0031 / 0804 / 1252 / 2022	6.2 / 0.9 / 6.2 / 1.1
8 F	0133 / 0904 / 1402 / 2124	6.5 / 0.7 / 6.4 / 0.8		23 Sa	0057 / 0839 / 1323 / 2056	6.1 / 1.1 / 6.1 / 1.3
9 Sa	0223 / 0955 / 1450 / 2216	6.2 / 1.1 / 6.1 / 1.2		24 Su	0131 / 0912 / 1359 / 2134	5.9 / 1.3 / 5.9 / 1.4
10 Su	0318 / 1051 / 1545 / 2318	5.9 / 1.5 / 5.8 / 1.5		25 M	0215 / 0952 / 1449 / 2216	5.7 / 1.5 / 5.8 / 1.6
11 M	0421 / 1157 / 1651	5.5 / 1.7 / 5.5		26 Tu	0314 / 1040 / 1550 / 2309	5.5 / 1.7 / 5.6 / 1.7
12 Tu	0027 / 0544 / 1309 / 1812	1.6 / 5.3 / 1.9 / 5.4		27 W	0430 / 1139 / 1705	5.4 / 1.8 / 5.6
13 W	0142 / 0713 / 1423 / 1928	1.6 / 5.4 / 1.8 / 5.6		28 Th	0017 / 0555 / 1250 / 1819	1.7 / 5.5 / 1.8 / 5.7
14 Th	0256 / 0816 / 1529 / 2025	1.5 / 5.6 / 1.6 / 5.8		29 F	0135 / 0657 / 1406 / 1923	1.5 / 5.7 / 1.6 / 6.0
15 F	0357 / 0901 / 1624 / 2110	1.3 / 5.8 / 1.4 / 6.0		30 Sa	0247 / 0812 / 1512 / 2018	1.2 / 6.0 / 1.3 / 6.3
				31 Su	0349 / 0847 / 1612 / 2108	1.0 / 6.3 / 1.1 / 6.5

JUNE

Day	Time	m		Day	Time	m
1 M	0444 / 0938 / 1705 / 2159	0.8 / 6.5 / 0.9 / 6.7		16 Tu	0518 / 1024 / 1734 / 2237	1.3 / 6.1 / 1.3 / 6.2
2 Tu ●	0537 / 1028 / 1757 / 2248	0.6 / 6.6 / 0.7 / 6.8		17 W ○	0554 / 1101 / 1814 / 2312	1.1 / 6.3 / 1.1 / 6.3
3 W	0629 / 1120 / 1848 / 2339	0.6 / 6.7 / 0.7 / 6.8		18 Th	0634 / 1133 / 1853 / 2343	1.0 / 6.3 / 1.0 / 6.2
4 Th	0721 / 1212 / 1938	0.6 / 6.7 / 0.6		19 F	0712 / 1205 / 1931	1.0 / 6.3 / 1.0
5 F	0031 / 0811 / 1300 / 2027	6.7 / 0.6 / 6.6 / 0.7		20 Sa	0014 / 0748 / 1238 / 2006	6.2 / 1.0 / 6.3 / 1.1
6 Sa	0120 / 0900 / 1345 / 2118	6.5 / 0.8 / 6.4 / 0.8		21 Su	0048 / 0822 / 1313 / 2042	6.2 / 1.1 / 6.3 / 1.1
7 Su	0208 / 0948 / 1430 / 2207	6.2 / 1.1 / 6.2 / 1.1		22 M	0126 / 0856 / 1352 / 2118	6.1 / 1.2 / 6.2 / 1.2
8 M	0257 / 1037 / 1519 / 2259	5.9 / 1.4 / 6.0 / 1.4		23 Tu	0211 / 0935 / 1437 / 2159	6.0 / 1.3 / 6.1 / 1.3
9 Tu	0353 / 1129 / 1617 / 2356	5.7 / 1.7 / 5.8 / 1.6		24 W	0304 / 1019 / 1532 / 2247	5.8 / 1.5 / 5.9 / 1.4
10 W	0459 / 1227 / 1723	5.4 / 1.9 / 5.6		25 Th	0409 / 1111 / 1634 / 2346	5.7 / 1.6 / 5.9 / 1.5
11 Th	0059 / 0617 / 1328 / 1836	1.7 / 5.4 / 2.0 / 5.6		26 F	0516 / 1214 / 1742	5.7 / 1.7 / 5.9
12 F	0206 / 0724 / 1436 / 1940	1.8 / 5.4 / 2.0 / 5.7		27 Sa	0056 / 0624 / 1326 / 1848	1.5 / 5.8 / 1.6 / 6.0
13 Sa	0311 / 0819 / 1535 / 2033	1.7 / 5.6 / 1.8 / 5.9		28 Su	0208 / 0727 / 1437 / 1949	1.4 / 5.9 / 1.5 / 6.1
14 Su	0402 / 0905 / 1620 / 2118	1.6 / 5.8 / 1.7 / 6.0		29 M	0317 / 0826 / 1542 / 2049	1.2 / 6.1 / 1.3 / 6.3
15 M	0442 / 0948 / 1658 / 2200	1.4 / 6.0 / 1.5 / 6.2		30 Tu	0420 / 0938 / 1644 / 2143	1.1 / 6.3 / 1.1 / 6.5

JULY

Day	Time	m		Day	Time	m
1 W ●	0520 / 1020 / 1742 / 2238	0.9 / 6.5 / 1.0 / 6.6		16 Th	0530 / 1037 / 1754 / 2247	1.2 / 6.2 / 1.2 / 6.2
2 Th	0621 / 1112 / 1839 / 2330	0.8 / 6.6 / 0.8 / 6.6		17 F ○	0614 / 1109 / 1836 / 2319	1.1 / 6.3 / 1.1 / 6.3
3 F	0716 / 1200 / 1933	0.8 / 6.7 / 0.7		18 Sa	0655 / 1143 / 1917 / 2354	1.0 / 6.5 / 1.0 / 6.3
4 Sa	0019 / 0806 / 1243 / 2022	6.6 / 0.8 / 6.6 / 0.7		19 Su	0733 / 1218 / 1954	1.0 / 6.5 / 0.9
5 Su	0104 / 0851 / 1326 / 2107	6.5 / 0.9 / 6.5 / 0.8		20 M	0032 / 0806 / 1257 / 2027	6.4 / 1.0 / 6.5 / 0.9
6 M	0147 / 0934 / 1406 / 2150	6.3 / 1.1 / 6.4 / 1.0		21 Tu	0114 / 0839 / 1338 / 2101	6.4 / 1.0 / 6.4 / 1.0
7 Tu	0232 / 1012 / 1451 / 2231	6.1 / 1.4 / 6.2 / 1.3		22 W	0159 / 0915 / 1420 / 2141	6.3 / 1.1 / 6.3 / 1.1
8 W	0319 / 1049 / 1541 / 2313	5.8 / 1.7 / 6.0 / 1.6		23 Th	0249 / 0954 / 1510 / 2224	6.1 / 1.3 / 6.2 / 1.2
9 Th	0414 / 1129 / 1637	5.6 / 1.9 / 5.8		24 F	0343 / 1044 / 1604 / 2316	6.0 / 1.5 / 6.0 / 1.4
10 F	0000 / 0518 / 1214 / 1742	1.8 / 5.4 / 2.2 / 5.6		25 Sa	0444 / 1142 / 1708	5.8 / 1.7 / 5.9
11 Sa	0052 / 0631 / 1310 / 1853	2.0 / 5.3 / 2.3 / 5.5		26 Su	0021 / 0551 / 1252 / 1818	1.6 / 5.7 / 1.8 / 5.8
12 Su	0157 / 0738 / 1420 / 1957	2.1 / 5.4 / 2.2 / 5.6		27 M	0137 / 0703 / 1411 / 1930	1.6 / 5.7 / 1.7 / 5.8
13 M	0304 / 0834 / 1529 / 2060	1.9 / 5.6 / 2.0 / 5.7		28 Tu	0256 / 0815 / 1527 / 2039	1.5 / 5.9 / 1.5 / 6.0
14 Tu	0359 / 0921 / 1623 / 2135	1.7 / 5.8 / 1.7 / 5.9		29 W	0409 / 0922 / 1634 / 2142	1.3 / 6.1 / 1.3 / 6.2
15 W	0445 / 1002 / 1709 / 2214	1.5 / 6.0 / 1.4 / 6.0		30 Th	0516 / 1017 / 1737 / 2235	1.1 / 6.4 / 1.1 / 6.4
				31 F ●	0619 / 1102 / 1835 / 2322	1.0 / 6.5 / 0.9 / 6.5

AUGUST

Day	Time	m		Day	Time	m
1 Sa	0713 / 1144 / 1927	0.9 / 6.7 / 0.8		16 Su	0638 / 1118 / 1902 / 2332	1.0 / 6.6 / 0.9 / 6.5
2 Su	0004 / 0759 / 1224 / 2011	6.6 / 0.9 / 6.7 / 0.7		17 M	0717 / 1156 / 1938	0.9 / 6.7 / 0.8
3 M	0043 / 0837 / 1302 / 2050	6.5 / 0.9 / 6.7 / 0.8		18 Tu	0011 / 0749 / 1235 / 2009	6.6 / 0.9 / 6.8 / 0.8
4 Tu	0123 / 0910 / 1340 / 2124	6.4 / 1.1 / 6.6 / 1.0		19 W	0055 / 0819 / 1317 / 2042	6.6 / 1.0 / 6.7 / 0.9
5 W	0201 / 0936 / 1419 / 2155	6.2 / 1.4 / 6.4 / 1.3		20 Th	0138 / 0854 / 1359 / 2119	6.5 / 1.2 / 6.6 / 0.9
6 Th	0243 / 1000 / 1501 / 2226	6.0 / 1.6 / 6.2 / 1.5		21 F	0226 / 0934 / 1446 / 2200	6.3 / 1.4 / 6.3 / 1.1
7 F	0328 / 1031 / 1554 / 2302	5.7 / 1.9 / 5.9 / 1.8		22 Sa	0317 / 1020 / 1538 / 2251	6.1 / 1.6 / 6.1 / 1.4
8 Sa	0421 / 1112 / 1645 / 2349	5.4 / 2.1 / 5.5 / 2.1		23 Su	0416 / 1115 / 1641 / 2357	5.8 / 1.7 / 5.8 / 1.7
9 Su	0529 / 1205 / 1757	5.2 / 2.4 / 5.3		24 M	0525 / 1229 / 1756	5.6 / 1.9 / 5.6
10 M	0046 / 0650 / 1313 / 1917	2.2 / 5.1 / 2.4 / 5.2		25 Tu	0120 / 0648 / 1357 / 1923	1.9 / 5.5 / 1.9 / 5.6
11 Tu	0158 / 0801 / 1437 / 2020	2.2 / 5.3 / 2.3 / 5.4		26 W	0249 / 0816 / 1521 / 2047	1.7 / 5.7 / 1.6 / 5.8
12 W	0317 / 0853 / 1550 / 2110	2.0 / 5.6 / 1.9 / 5.6		27 Th	0406 / 0922 / 1633 / 2148	1.5 / 6.0 / 1.3 / 6.1
13 Th	0416 / 0934 / 1644 / 2148	1.7 / 5.9 / 1.5 / 5.9		28 F	0515 / 1009 / 1734 / 2233	1.2 / 6.3 / 1.0 / 6.4
14 F	0506 / 1009 / 1733 / 2220	1.3 / 6.2 / 1.2 / 6.1		29 Sa	0614 / 1048 / 1829 / 2309	1.0 / 6.6 / 0.8 / 6.5
15 Sa ○	0554 / 1042 / 1818 / 2254	1.1 / 6.4 / 1.0 / 6.4		30 Su	0702 / 1125 / 1914 / 2344	0.9 / 6.7 / 0.8 / 6.5
				31 M	0741 / 1201 / 1952	0.9 / 6.8 / 0.8

Figure 60 Dover Tide tables May to August.

Examine Fig. 60 again, and look at the ranges of the Dover tides for the period 1 to 15 May. The highest range of tide occurs on the 5th and 6th so these must be spring tides; note that the entry for the 4th is marked with a small closed circle, denoting the day of the new moon (ie the sun and moon are in line at that time). Observe that the 19 May, some fifteen days later, is marked with a small open circle denoting the date of the full moon and the approximate time of the next spring. The smallest range, the neap tide, between these two dates occurs on the morning of the 12th, being only 3.4m (5.3–1.9m).

There is always a slight delay – one to two days – between the date of a new or full moon and the associated spring tide. The tide tables give the times of high and low water in GMT, now universally quoted as universal time (UT). Notice that at the top of the page in Fig. 60 a reference is made to time zone UT (GMT), and a warning given regarding the UK's change to British summer time (BST). One hour is added to GMT to produce BST on 21 March, and it is taken off again on 22 October.

GMT and Time Zones

The sun is used, the world over, to calculate local mean time (LMT). At every position on earth, noon is that instant when the sun crosses the meridian: it reaches its greatest height at this time and is said to be at its culmination. It appears to move across our sky from east to west, completing one 360° revolution every twenty-four hours, that is, at an average rotational speed of 15° per hour. A consequence of this is that at local noon – 1200

hours – at Greenwich, the LMT at a position 90° east of Greenwich is 1800 hours of that same day (as shown in Fig. 61); ie noon at this second position occurred six hours earlier as the sun crossed that position's meridian on its westward journey. At this same instant the LMT at a position 90° west of Greenwich would be 0600 hours of that same date, ie local noon here would occur some six hours later when the sun would be overhead. These details are shown on Fig. 61, together with the very interesting time phenomena experienced at that position on the earth's surface which is 180° E/W of Greenwich, this position being the 'international date line'(IDL).

At the instant of local noon at Greenwich, the LMT at the position immediately east of the IDL is 0001 hours of the same day, whereas at the position immediately west of the IDL the LMT is 2359 of that same day.

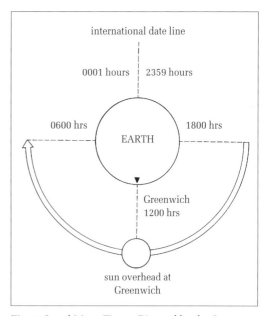

Fig 61 Local Mean Time – Dictated by the Sun.

Imagine now a vessel about to cross the IDL in a westerly direction at 0001 on a Monday morning: as it crosses the IDL, the LMT would become 0001 on Tuesday, so that virtually the whole of Monday would simply not exist to that crossing vessel. A vessel about to cross the IDL in an easterly direction at 2359 LMT on Monday would have two Mondays that week because as she crossed the IDL her LMT would revert to 0001 of that same Monday. Unfortunately, magic though it may seem, crossing the IDL in an easterly direction would not allow you to cancel a mistake made earlier in that same day.

Local time, a very human and necessary facility, produces a requirement for an internationally recognized time reference, and this is the LMT at Greenwich, until recently called Greenwich mean time (GMT) but now universally referred to as universal time (UT). Both LMT and GMT are kept on board all sea-going vessels. Where LMT governs the normal daily activity of those on board, GMT on the other hand is required in order that those people concerned with the navigation of the ship can reference their many technical data tables, which are always referenced to GMT.

In order to relate LMT to GMT, the world is divided into twenty-four time zones, each one occupying 15° of longitude. Time zone 0 is centred on the Greenwich meridian covering the area 7.5°E to 7.5°W of Greenwich. The zones east of Greenwich are numbered one through to twelve with a negative prefix, −1 to −12; the zones west of Greenwich are numbered +1 to +12. Conventionally the + prefix is not normally shown on documentation. Time zone 12 straddles the IDL and is consequently two half time zones, each covering 7.5° of longitude and signed −12 and +12. (The time-zoned earth is shown at Fig. 62.) Local time,

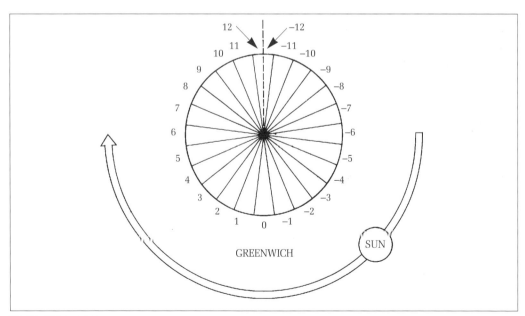

Fig 62 Time zones.

TIME ZONE -0100
(French Standard Time)
Subtract 1 hour for GMT
For French Summer Time add
ONE hour in non-shaded area

FRANCE, NORTH COAST — DIEPPE

Lat 49'56' N Long 1'05' E

TIMES AND HEIGHTS OF HIGH AND LOW WATERS

MAY

Day	Time	m		Day	Time	m
1 F	0342 / 0921 / 1612 / 2148	1.5 / 8.3 / 1.2 / 8.5		16 Sa	0455 / 1026 / 1716 / 2245	1.5 / 8.3 / 1.4 / 8.3
2 Sa	0439 / 1013 / 1705 / 2237	1.0 / 8.9 / 0.7 / 9.0		17 Su	0536 / 1103 / 1754 / 2321	1.3 / 8.5 / 1.2 / 8.5
3 Su ●	0530 / 1100 / 1754 / 2323	0.6 / 9.3 / 0.4 / 9.4		18 M ○	0612 / 1137 / 1830 / 2354	1.1 / 8.6 / 1.1 / 8.7
4 M	0618 / 1146 / 1841	0.3 / 9.6 / 0.3		19 Tu	0646 / 1210 / 1902	1.1 / 8.7 / 1.1
5 Tu	0008 / 0703 / 1231 / 1925	9.6 / 0.3 / 9.7 / 0.3		20 W	0027 / 0718 / 1243 / 1934	8.8 / 1.0 / 8.7 / 1.1
6 W	0054 / 0746 / 1316 / 2009	9.7 / 0.3 / 9.6 / 0.4		21 Th	0100 / 0749 / 1316 / 2005	8.7 / 1.1 / 8.6 / 1.2
7 Th	0138 / 0831 / 1402 / 2052	9.5 / 0.5 / 9.3 / 0.7		22 F	0132 / 0821 / 1348 / 2037	8.6 / 1.2 / 8.6 / 1.4
8 F	0223 / 0912 / 1445 / 2134	9.1 / 0.8 / 8.8 / 1.1		23 Sa	0206 / 0852 / 1422 / 2109	8.5 / 1.4 / 8.1 / 1.6
9 Sa	0308 / 0956 / 1532 / 2220	8.6 / 1.3 / 8.1 / 1.6		24 Su	0240 / 0926 / 1459 / 2145	8.2 / 1.6 / 1.8 / 1.9
10 Su	0358 / 1047 / 1628 / 2317	8.0 / 1.8 / 7.5 / 2.1		25 M	0319 / 1006 / 1541 / 2229	7.9 / 1.9 / 7.5 / 2.1
11 M	0500 / 1151 / 1737	7.5 / 2.2 / 7.0		26 Tu	0407 / 1054 / 1634 / 2323	7.6 / 2.1 / 7.2 / 2.3
12 Tu	0029 / 0618 / 1309 / 1900	2.5 / 7.3 / 2.3 / 7.0		27 W	0507 / 1156 / 1743	7.4 / 2.2 / 7.1
13 W	0152 / 0741 / 1432 / 2016	2.4 / 7.4 / 2.2 / 7.2		28 Th	0033 / 0622 / 1312 / 1901	2.3 / 7.5 / 2.0 / 7.4
14 Th	0307 / 0848 / 1539 / 2116	2.0 / 7.6 / 1.9 / 7.6		29 F	0151 / 0737 / 1427 / 2012	2.0 / 7.8 / 1.7 / 7.9
15 F	0406 / 0942 / 1632 / 2205	1.8 / 8.0 / 1.6 / 8.0		30 Sa	0302 / 0843 / 1533 / 2112	1.5 / 8.3 / 1.2 / 8.5
				31 Su	0403 / 0941 / 1632 / 2207	1.1 / 8.8 / 0.9 / 8.9

JUNE

Day	Time	m		Day	Time	m
1 M	0459 / 1033 / 1727 / 2258	0.7 / 9.2 / 0.6 / 9.3		16 Tu	0540 / 1107 / 1759 / 2326	1.4 / 8.3 / 1.4 / 8.5
2 Tu ●	0553 / 1123 / 1819 / 2348	0.5 / 9.4 / 0.5 / 9.5		17 W ○	0619 / 1145 / 1838	1.3 / 8.4 / 1.3
3 W	0643 / 1211 / 1907	0.4 / 9.4 / 0.5		18 Th	0003 / 0655 / 1221 / 1913	8.6 / 1.2 / 8.5 / 1.2
4 Th	0036 / 0731 / 1301 / 1954	9.5 / 0.4 / 9.3 / 0.6		19 F	0038 / 0730 / 1257 / 1947	8.7 / 1.1 / 8.5 / 1.2
5 F	0125 / 0817 / 1348 / 2040	9.4 / 0.6 / 9.1 / 0.8		20 Sa	0115 / 0805 / 1333 / 2022	8.7 / 1.1 / 8.5 / 1.3
6 Sa	0211 / 0901 / 1434 / 2122	9.1 / 0.9 / 8.7 / 1.2		21 Su	0151 / 0841 / 1409 / 2057	8.6 / 1.2 / 8.3 / 1.4
7 Su	0256 / 0944 / 1519 / 2206	8.7 / 1.2 / 8.2 / 1.5		22 M	0228 / 0915 / 1447 / 2134	8.5 / 1.3 / 8.1 / 1.5
8 M	0343 / 1031 / 1608 / 2256	8.2 / 1.6 / 7.7 / 2.0		23 Tu	0307 / 0953 / 1528 / 2215	8.3 / 1.5 / 7.9 / 1.7
9 Tu	0435 / 1123 / 1706 / 2354	7.8 / 2.0 / 7.3 / 2.3		24 W	0351 / 1038 / 1617 / 2303	8.1 / 1.6 / 7.7 / 1.8
10 W	0537 / 1226 / 1812	7.5 / 2.2 / 7.1		25 Th	0443 / 1132 / 1714	7.9 / 1.8 / 7.6
11 Th	0101 / 0648 / 1337 / 1924	2.4 / 7.3 / 2.3 / 7.1		26 F	0002 / 0547 / 1235 / 1822	1.8 / 7.8 / 1.8 / 7.7
12 F	0212 / 0757 / 1445 / 2027	2.3 / 7.4 / 2.2 / 7.4		27 Sa	0111 / 0658 / 1348 / 1934	1.8 / 8.0 / 1.6 / 7.9
13 Sa	0316 / 0855 / 1544 / 2121	2.1 / 7.6 / 2.0 / 7.7		28 Su	0224 / 0808 / 1459 / 2041	1.6 / 8.2 / 1.4 / 8.3
14 Su	0411 / 0945 / 1635 / 2208	1.9 / 7.9 / 1.9 / 8.0		29 M	0332 / 0912 / 1605 / 2143	1.3 / 8.6 / 1.1 / 8.7
15 M	0458 / 1029 / 1719 / 2249	1.6 / 8.1 / 1.6 / 8.3		30 Tu	0436 / 1012 / 1706 / 2240	1.0 / 8.8 / 0.9 / 9.0

JULY

Day	Time	M		Day	Time	M
1 W ●	0536 / 1107 / 1804 / 2333	0.8 / 9.0 / 0.8 / 9.2		16 Th ○	0554 / 1122 / 1816 / 2341	1.4 / 8.2 / 1.4 / 8.5
2 Th	0631 / 1159 / 1856	0.6 / 9.1 / 0.7		17 F	0636 / 1201 / 1854	1.2 / 8.5 / 1.2
3 F	0025 / 0720 / 1250 / 1944	9.3 / 0.6 / 9.1 / 0.7		18 Sa	0021 / 0713 / 1239 / 1932	8.8 / 1.1 / 8.6 / 1.1
4 Sa	0114 / 0806 / 1336 / 2028	9.3 / 0.7 / 9.0 / 0.8		19 Su	0059 / 0749 / 1318 / 2008	8.9 / 1.0 / 8.7 / 1.0
5 Su	0159 / 0848 / 1420 / 2108	9.3 / 0.8 / 8.7 / 1.1		20 M	0136 / 0827 / 1356 / 2045	8.9 / 0.9 / 8.7 / 1.0
6 M	0240 / 0927 / 1500 / 2147	8.8 / 1.1 / 8.4 / 1.4		21 Tu	0215 / 0902 / 1434 / 2120	8.9 / 1.0 / 8.6 / 1.1
7 Tu	0321 / 1007 / 1542 / 2228	8.5 / 1.4 / 8.0 / 1.7		22 W	0252 / 0939 / 1513 / 2159	8.8 / 1.1 / 8.4 / 1.3
8 W	0404 / 1050 / 1627 / 2312	8.1 / 1.8 / 7.6 / 2.0		23 Th	0333 / 1020 / 1556 / 2242	8.6 / 1.2 / 8.2 / 1.4
9 Th	0451 / 1137 / 1718	7.7 / 2.1 / 7.2		24 F	0419 / 1106 / 1646 / 2334	8.3 / 1.5 / 8.0 / 1.6
10 F	0005 / 0547 / 1235 / 1820	2.3 / 7.3 / 2.3 / 7.0		25 Sa	0515 / 1203 / 1747	8.1 / 1.6 / 7.8
11 Sa	0107 / 0654 / 1343 / 1928	2.5 / 7.2 / 2.5 / 7.0		26 Su	0037 / 0624 / 1314 / 1901	1.8 / 7.9 / 1.8 / 7.8
12 Su	0217 / 0801 / 1450 / 2032	2.5 / 7.2 / 2.4 / 7.2		27 M	0154 / 0741 / 1433 / 2018	1.8 / 7.9 / 1.7 / 8.0
13 M	0322 / 0902 / 1552 / 2129	2.3 / 7.4 / 2.2 / 7.6		28 Tu	0311 / 0854 / 1548 / 2129	1.6 / 8.1 / 1.5 / 8.4
14 Tu	0420 / 0954 / 1645 / 2218	2.0 / 7.6 / 1.9 / 7.9		29 W	0423 / 1002 / 1656 / 2231	1.3 / 8.5 / 1.2 / 8.7
15 W	0509 / 1040 / 1732 / 2301	1.7 / 7.9 / 1.6 / 8.3		30 Th	0527 / 1059 / 1755 / 2326	1.0 / 8.7 / 1.0 / 9.0
				31 F ●	0622 / 1151 / 1847	0.8 / 8.9 / 0.8

AUGUST

Day	Time	M		Day	Time	M
1 Sa	0016 / 0710 / 1238 / 1932	9.2 / 0.7 / 9.0 / 0.8		16 Su	0001 / 0654 / 1221 / 1913	8.9 / 0.9 / 8.8 / 0.9
2 Su	0101 / 0752 / 1322 / 2013	9.3 / 0.6 / 9.0 / 0.8		17 M	0040 / 0732 / 1300 / 1950	9.2 / 0.7 / 9.0 / 0.7
3 M	0141 / 0832 / 1401 / 2049	9.2 / 0.7 / 8.8 / 0.9		18 Tu	0119 / 0809 / 1337 / 2028	9.3 / 0.6 / 9.1 / 0.7
4 Tu	0218 / 0905 / 1436 / 2121	9.0 / 0.9 / 8.6 / 1.2		19 W	0157 / 0846 / 1415 / 2103	9.3 / 0.6 / 9.0 / 0.7
5 W	0253 / 0939 / 1511 / 2155	8.7 / 1.2 / 8.3 / 1.5		20 Th	0234 / 0920 / 1453 / 2140	9.2 / 0.7 / 8.9 / 0.9
6 Th	0328 / 1013 / 1547 / 2231	8.3 / 1.5 / 7.8 / 1.8		21 F	0313 / 1000 / 1533 / 2220	8.9 / 1.0 / 8.6 / 1.2
7 F	0406 / 1050 / 1627 / 2311	7.8 / 2.0 / 7.4 / 2.2		22 Sa	0356 / 1043 / 1620 / 2307	8.5 / 1.4 / 8.2 / 1.6
8 Sa	0450 / 1135 / 1716	7.4 / 2.4 / 7.0		23 Su	0448 / 1136 / 1720	8.0 / 1.8 / 7.8
9 Su	0003 / 0546 / 1235 / 1821	2.6 / 6.9 / 2.7 / 6.8		24 M	0010 / 0557 / 1249 / 1839	1.6 / 7.6 / 2.1 / 7.5
10 M	0111 / 0659 / 1350 / 1937	2.8 / 6.8 / 2.8 / 6.8		25 Tu	0133 / 0724 / 1418 / 2007	2.1 / 7.5 / 2.1 / 7.7
11 Tu	0230 / 0815 / 1507 / 2050	2.7 / 6.9 / 2.6 / 7.2		26 W	0302 / 0848 / 1542 / 2125	1.9 / 7.8 / 1.8 / 8.1
12 W	0342 / 0921 / 1613 / 2149	2.3 / 7.3 / 2.2 / 7.6		27 Th	0418 / 0957 / 1652 / 2227	1.5 / 8.2 / 1.4 / 8.6
13 Th	0441 / 1015 / 1707 / 2238	1.9 / 7.7 / 1.8 / 8.1		28 F ●	0521 / 1054 / 1747 / 2318	1.1 / 8.6 / 1.1 / 9.0
14 F ○	0531 / 1100 / 1753 / 2321	1.5 / 8.2 / 1.4 / 8.6		29 Sa	0612 / 1140 / 1835	0.9 / 8.9 / 0.9
15 Sa	0615 / 1141 / 1835	1.1 / 8.6 / 1.1		30 Su	0002 / 0656 / 1223 / 1915	9.2 / 0.7 / 9.0 / 0.7
				31 M	0041 / 0733 / 1301 / 1950	9.3 / 0.6 / 9.1 / 0.7

Fig 63 Dieppe tide tables.

anywhere on the earth, is now easily converted to the international time reference by making an algebraic sum of local time and UT. Therefore a location 090° east is in time zone –6, and from our example above, its local time of 1800 is the equivalent of 1200 UT, noon at Greenwich; thus: 1800(LMT)–6(time zone)=1200(UT).

The European yachtsman travelling between UK (TZ = 0000) and continental ports (TZ = –0100) will obtain tidal information from the relevant pages of the almanac; the time data on these pages is given in LMT, consequently the data for French ports (TZ=–0100) would need to be converted to UK times (TZ=0000) when passage planning. For example, using Fig. 63, the tidal table for Dieppe (TZ=–0100), the LMT of HW at Dieppe on the morning of Monday 11 May is 1500; the TZ of –0100 gives the GMT of the HW at Dieppe of 0500–0100 = 0400 GMT.

The Twelfths Rule

This rule assumes a sinusoidal shape for the rise and fall of the tide; it also assumes a twelve-hour period from low water through high water and back to the next low water (Fig. 64). It further assumes that the rise in sea level follows the sequence below:

During the first and the sixth hour after LW, it rises by one-twelfth of the range.

During the second and the fifth hour after LW, it rises by two-twelfths of the range.

During the third and the fourth hour after LW, it rises by three-twelfths of the range.

The final assumption is that the fall of tide from HW to LW follows the same hourly pattern as the rise. For example,

assuming a tidal range of 3.5m, calculate the rise of tide 3½ hours after LW, using the twelfths rule:

During first hour rise= 1/12*3.5 =0.29m.

During second hour rise= 2/12*3.5 =0.58m.

During third hour rise= 3/12*3.5 =0.87m.

During last half-hour rise= ½*3/12*3.5 =0.44m.

Total rise in 3½ hours = 2.18m.

Examination of an actual port's tidal curve will usually show only a passing resemblance to the fabricated twelfths curve: in other words, the shape and timing of the twelfths rule tide can bear little relationship to the actual tide you may be experiencing. Nonetheless it is a useful check on your tidal predictions.

When you have completed your tidal calculations using the correct tide curve and data for your port of interest, see later, the twelfths rule should produce a reasonable comparison, and if it does not, you should check your calculations.

I have known mariners who do their tidal predictions using only the twelfths rule, but all of them must leave a very large safety margin for arrival and departure times, at least the ones who are still afloat must do so.

Tidal Curves

The twelfths rule assumes a sinusoidal rise and fall of tide, but in fact a quick skim through the tidal curves of individual ports given in a nautical almanac will show that this is only an approximation. For instance Fig. 65 is a copy of the tidal curve for Portsmouth; it shows the predicted characteristic curve for the full tidal period of LW through HW and back to LW, and is most certainly far from sinu-

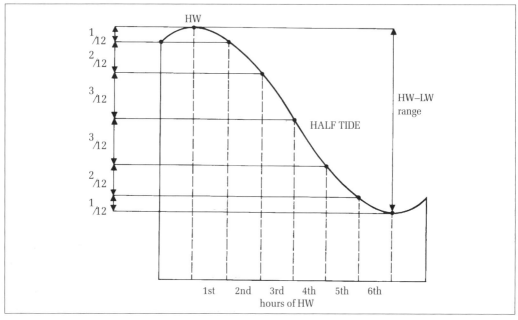

Fig 64 The Twelfths Rule. Assumptions are sinusoidal tide of 12 hours duration HW to HW.

soidal. Notice also from Fig. 65 how the curve for a neap tide, the dotted rise and fall, differs from the full line of the spring tide. The boxes along the base line of the curve allow the navigator to enter the time of HW, and the hourly intervals before and after this time. Fig. 66, a copy

of Dover's tidal curve, is also far from the idealized curve of the twelfths rule.

The navigator wishing to enter a particular port obtains from the nautical tide

Fig 65 Portsmouth's tidal curve.

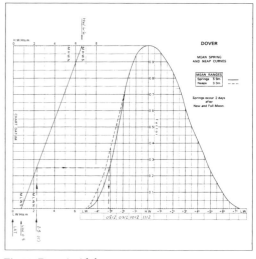

Fig 66 Dover's tidal curve.

tables the time and height of low water for the date in question, and the LAT from the appropriate chart. He then calculates the depth of water for the time in question, compares this depth with his yacht's draught, plus a safety margin to determine whether entry or departure can be made at any state of the tide. In situations where the depth at low water is too low, the mariner must wait for a rise of tide which will satisfy the following: rise of tide + LW + LAT >= yacht's draught + safety.

For example, let us assume that LW = 0.8m, LAT = 0.3m, draught = 2.0m and safety = 0.5m. Thus at low water the depth would be 1.1m, and entry would need to be delayed until the tide had risen at least 1.4m above its low-water level.

As we have seen, the tide tables for a particular port give the heights and times of high and low water only, and the mariner must turn to the tidal curve of the port in question in order to calculate the earliest time after LW, when the depth will have risen to equal his requirements of draught + safety. The tidal curve for Dover, Fig. 66, has two parts: the first part is a graph, and you may wish to turn the diagram through 90° to see the graph in proper perspective. The left-hand scale (a) of Fig. 66 allows the height of HW to be entered; the right-hand scale (b) is for the height of LW to be entered. Remember that HW and LW are measured above LAT; this means that the baseline of this graph is at LAT, therefore soundings will be below the baseline and drying heights will be above it. As an example, values for the morning tide at Dover on the 2 July, obtained from Fig. 60 are: LW − 0621(0.8m); HW − 1112(6.6m). These values have been entered on Fig. 66(a), and are shown joined by a fine straight line.

Turn the diagram back through 90° and note part (b): this is the tidal curve. At the bottom of the curve are the baseline time boxes. The time of HW (1112(GMT)) has been entered in the centre box, as have the times of the periods both before and after HW in the appropriate side boxes.

Now, let us assume that a yacht requiring 2.5m is to enter Dover on the morning of this same day, 2 July, and berth in a location where the LAT is only 0.2m. The minimum height of tide required is 2.3m (draught–LAT), but since the height of tide at LW is only 0.8m, clearly the yacht cannot berth at LW.

The height of tide required, 2.3m, has been entered on Fig. 66 at (c), and extended to meet the line joining together the tides HW/LW at point (d), the line is then projected to cut the tidal curve on a rising neap tide (point e), and finally the time at which this height will occur is read off from the baseline time boxes. The centre of the time box is three hours before HW, in this case 0812 GMT, and interpolation within the box gives the earliest time to enter as 0752GMT.

In the above example we started by deducing the minimum depth required, the tide height scale was entered with this value, and, by reference to the tidal curve, we obtained the time at which the required minimum depth would be reached.

This working predicts the time of the absolute minimum depth required, but the prudent mariner will also allow for sea chop and wave height, the date the LAT was recorded, high pressure weather and the fact that our calculations are only predictions.

All the above factors could, and Murphy says usually would, give less depth than calculated. In this example we have entered on a rising tide and so touching a soft bottom may be only a momentary embarrassment: however, it is sometimes necessary to enter or leave on a falling tide, in which circumstance it is good practice to add one metre or more, depending on the conditions prevailing, to the minimum depth required.

On some occasions we would use the above tide graph and curve in the reverse order to the above; for example, when planning an arrival time having decided to make for a particular port during a cruise – for our purposes here, let the port be Dover with the tidal details as used in Fig. 66. The skipper calculates the estimated time of arrival and compares it with the time of HW at the port in question.

Let us assume a yacht drawing 2.5m desiring a safety margin of 0.5m – that is, a required minimum depth of 3m on arrival – is 10 n. miles off the port, making 5 knots, with the port dead ahead; the ETA is therefore two hours away. A reference to the tidal curve of the port will allow comparison of the ETA with the time of HW. Assuming that the yacht will arrive 3.5 hours before HW on a rising spring tide, in this case 0742 GMT, we enter the appropriate time box at the base of the curve, extend the entry point to cut the tidal curve, project this point on the curve down to the line on the graph joining HW to LW, and hence read the height of tide at our ETA. From Fig. 66, the height of tide at 0742 is only 1.6m. With this information the skipper then checks the chart for LAT values.

In this last example of using the curve and graph, the mariner has useful tide information some two hours before arrival, allowing the skipper to plan, in detail, how the port entry will be made.

The importance of a mariner's ability to use tide tables, graphs and curves correctly cannot be overstated; going aground is not funny, it is at the least embarrassing, and can be extremely dangerous. Many yachts have spent some unscheduled LW time parked on the putty, and have fortunately lived to float again on the next flood. It is worthwhile at this point to ponder the mental pictures running through a skipper's mind when the yacht is fast aground on a falling tide: moving the crew around the boat, reversing the engine, sitting out on the boom, throwing the crew overboard, swearing, praying – but all to no avail because the yacht is slowly but definitely heeling over (assuming she is a single keel). She may eventually lie on her side.

In a 'worst case' scenario, the reader may recall that this ebbing tide is associated with the biggest spring range of the year; also, since this grounding is an unplanned event, the skipper has no means of knowing what she will lie on: sharp, piercing rock? An uneven, rock-strewn sea bed? An old broken anchor fluke? If she is damaged as she settles, will she fill with water on the next flood?

None of these questions need arise because tidal planning, good chartwork, observation and the practice of good seamanship mean that you may be an audience to such tales, but you should never be the one relating them.

7

SECONDARY PORTS

Secondary ports and harbours may be considered of lesser importance to large commercial shipping concerns than are the big standard ports such as Dover, because they are generally smaller, but almost therefore by definition they are of greater importance to the yachtsman. Tide predictions for secondary ports are printed in the nautical almanacs but are presented in a different way to the data for standard ports, where the heights and times of high and low water are printed for each tide of each day of the year.

For tide-prediction purposes all secondary ports are related to a nearby standard port whose characteristic tidal curve is assumed to have the same shape as that secondary port. The times and heights of high and low waters usually differ, however, and it is these differences which are published. Fig. 67a is a copy of the tide predictions for the secondary port of Yarmouth and Fig. 67b the tide table for the associated standard port, Portsmouth.

From the details of Fig. 67a, for Yarmouth, general tidal information is given first:

Yarmouth is on the Isle of Wight.
Latitude and longitude are given.
Important charts are listed.

Mean water times are given (i) for springs 50 mins before, and 1 hr 50 mins after mean double HW times at the port of Dover (–0050,+0150); (ii) for neap tides 20 mins later.
The mean water level (ML) is 2.0m.
Yarmouth is in time zone zero.
The standard port to which Yarmouth tidal differences apply is Portsmouth.

Below this general list of information the time and height differences are listed for spring and neap tides between Portsmouth and Yarmouth.

Time Differences

First of all we must accept that spring tides tend to occur at a place at the same time of day; for example, spring tides at Portsmouth tend to occur at noon and midnight. The first four columns are the time differences, and are read in the following way:
The first column entry under the heading 'High Water':
When the time of HW at Portsmouth is 0000 or 1200 UT, ie MHWS, HW at Yarmouth occurs 1 hr 5 mins earlier (–0105).

```
244      Harbour, Tidal And Coastal Information

YARMOUTH                              10-2-20
Isle of Wight 50°42'.39N 01°29'.97W

CHARTS
  AC 2021, 2040; Stanford 11; Imray C3, Y20; OS 196
TIDES
  Sp −0050, +0150, Np +0020 Dover; ML 2.0 (GMT)

Standard Port PORTSMOUTH (←→)

Times                         Height (metres)
  High Water   Low Water    MHWS MHWN MLWN MLWS
  0000  0600   0500  1100    4.7   3.8   1.9   0.8
  1200  1800   1700  2300
Differences YARMOUTH
  −0105 +0005  −0025 −0030   −1.6  −1.3  −0.4   0.0
FRESHWATER
  −0210 +0025  −0040 −0020   −2.1  −1.5  −0.4   0.0
TOTLAND BAY
  −0130 −0045  −0040 −0040   −2.0  −1.5  −0.5  −0.1
NOTE: Double HWs occur at or near springs; at other times
there is a stand which lasts about two hrs. Predictions refer
to the first HW when there are two. At other times they refer
to the middle of the stand. See 10-2-13.

SHELTER
Good shelter from all directions of wind and sea, but swell
enters if wind strong N/NE. Hbr gets very full, and may be
closed. 12 Orange Visitors' Bys outside (see charlist). Boats
over 15m LOA, 4m beam or 2.4m draft should give notice of
arrival to tel. Hbr dredged 2m from ent to bridge.
NAVIGATION
WPT 50°42'.60N 01°29'.93W, 008°/188° from/to front ldg lt,
0.28M. Dangers on appr are Black Rock and shallow water to
the N of the bkwtr. E of the pier there are 3 large buoys, the E
and W of which are Fl Y 5s. Beware ferries. The road bridge
across the R Yar opens for access to moorings or the BY up-
river, (May-Sept) 0800, 0900, 1000, 1200, 1400, 1600, 1730,
1830, 2000 (LT); on request (Oct-May). Speed limit 4kn in hbr
and apprs from pier hd, .l, prohib in hbr and above R Yar
bridge.
```

Fig 67(a) Tide differences Yarmouth.

The second column entry under 'High Water':

When the time of HW at Portsmouth is 0600 or 1800 UT, ie MHWN, HW at Yarmouth occurs 5 mins later (+0005).

These time differences are quoted for mean spring (4.7m) and neap HWs (3.8m) at Portsmouth.

The HW differences in height are quoted under the heading MHWS and MHWN respectively. They are −1.6m at MHWS and −1.3m at MHWN. A good way to interpret this information is to say that when HW Portsmouth is 3.8m, HW at Yarmouth will be (4.7−1.6) 3.1m, whereas

a Portsmouth tide of 3.8m will result in a HW Yarmouth of (3.8−1.3) 2.5m.

To calculate the time difference for high-water heights other than MHWS and MHWN, the mariner reproduces the above information in the form of a graph. Points a,b,c and d on Fig. 68a represent the spring and neap high-water time differences plotted on a 24-hour baseline, by joining all these points as illustrated. The resultant graph allows the reader to extract the time difference between HW Yarmouth and HW Portsmouth for times of high water Portsmouth. For example, the tide table for Portsmouth on Monday 8 June gives HW at 1602 height 4.3m; entering Fig. 68a with this value, as shown, gives a predicted time difference of −0017 mins. That is, for a Portsmouth HW at 1602 GMT, HW Yarmouth occurs at 1545 GMT (1602−17=1545).

Fig. 69 is a simple 'folded back' version of Fig. 68a, and this much simpler graph is possible because the tide time difference at 0000 and 1200 is the same value, ie −0105, and the time difference at 0600 and 1800 is also the same value, ie 0005; the HW differences tabulated in Fig. 68a are extracted from this folded version of the full graph.

The third and fourth columns of Fig. 67a give the Yarmouth–Portsmouth time differences for LW:

When the time of LW Portsmouth occurs at 0500 or at 1700, the Yarmouth low water occurs 25 mins earlier (−0025). When LW Portsmouth occurs at 1100 or at 2300, the Yarmouth low water is 30 mins earlier (−0030).

These four low-water values are points a,b,c and d on Fig. 68b, and joining these points together forms a 24-hour, low-water graph which is used in exactly the same way as the HW graph dealt with

ENGLAND, SOUTH COAST — PORTSMOUTH

Lat 50°48′ N   Long   1°07′ W

TIMES AND HEIGHTS OF HIGH AND LOW WATERS

TIME ZONE UT (GMT)
Summer Time add ONE hour in non-shaded area

MAY

Day	Time	m	Time	m	Time	m	Time	m
1 F	0147	1.2	0845	4.2	1410	0.8	2121	4.6
2 Sa	0238	0.8	0936	4.6	1501	0.5	2210	4.8
3 Su	0325	0.6	1027	4.8	1551	0.4	2259	5.0
4 M	0412	0.4	1117	4.9	1640	0.3	2345	5.1
5 Tu	0459	0.3	1207	5.0	1726	0.4		
6 W	0031	5.0	0546	0.4	1255	4.9	1809	0.5
7 Th	0115	4.9	0629	0.4	1344	4.8	1851	0.8
8 F	0157	4.7	0715	0.7	1434	4.6	1936	1.1
9 Sa	0243	4.5	0804	1.0	1527	4.4	2029	1.4
10 Su	0336	4.2	0902	1.3	1632	4.2	2135	1.7
11 M	0441	3.9	1012	1.5	1742	4.0	2256	1.8
12 Tu	0601	3.7	1133	1.6	1901	3.9		
13 W	0019	1.8	0722	3.7	1249	1.5	2008	4.0
14 Th	0127	1.6	0828	3.9	1348	1.3	2100	4.2
15 F	0217	1.3	0918	4.0	1432	1.2	2142	4.3
16 Sa	0257	1.1	0958	4.2	1511	1.0	2218	4.5
17 Su	0331	1.0	1034	4.4	1547	1.0	2251	4.5
18 M	0406	0.9	1109	4.5	1622	0.9	2323	4.6
19 Tu	0438	0.9	1144	4.4	1657	0.9	2356	4.5
20 W	0511	0.8	1219	4.4	1729	0.9		
21 Th	0027	4.4	0542	0.8	1255	4.3	1759	1.0
22 F	0059	4.3	0614	0.9	1330	4.2	1830	1.1
23 Sa	0133	4.2	0647	0.9	1405	4.1	1904	1.2
24 Su	0209	4.1	0723	1.1	1446	4.1	1945	1.4
25 M	0251	4.0	0807	1.2	1632	4.0	2034	1.5
26 Tu	0341	3.9	0900	1.3	1629	4.0	2138	1.6
27 W	0443	3.8	1007	1.4	1736	4.0	2252	1.6
28 Th	0556	3.8	1122	1.3	1846	4.1		
29 F	0006	1.5	0706	4.0	1234	1.2	1949	4.3
30 Sa	0109	1.2	0810	4.3	1336	0.8	2047	4.6
31 Su	0206	0.9	0908	4.5	1434	0.7	2141	4.8

JUNE

Day	Time	m	Time	m	Time	m	Time	m
1 M	0259	0.7	1003	4.7	1527	0.6	2233	4.9
2 Tu	0351	0.5	1058	4.8	1617	0.5	2324	5.0
3 W	0441	0.4	1151	4.9	1706	0.6		
4 Th	0011	4.9	0527	0.4	1241	4.8	1750	0.7
5 F	0056	4.8	0613	0.6	1330	4.7	1833	0.9
6 Sa	0139	4.7	0657	0.8	1418	4.6	1919	1.2
7 Su	0223	4.5	0745	1.2	1508	4.5	2010	1.4
8 M	0312	4.3	0837	1.2	1602	4.3	2109	1.7
9 Tu	0409	4.0	0938	1.5	1704	4.2	2216	1.8
10 W	0517	3.9	1046	1.6	1811	4.1	2330	1.8
11 Th	0630	3.8	1159	1.6	1917	4.0		
12 F	0039	1.7	0742	3.8	1304	1.5	2015	4.1
13 Sa	0137	1.5	0840	3.9	1357	1.4	2019	4.2
14 Su	0223	1.3	0928	4.1	1441	1.2	2143	4.3
15 M	0303	1.2	1008	4.3	1521	1.1	2220	4.4
16 Tu	0340	1.1	1046	4.3	1556	1.2	2256	4.5
17 W	0415	1.0	1122	4.4	1632	1.1	2330	4.5
18 Th	0449	1.0	1157	4.3	1706	1.1		
19 F	0004	4.4	0523	0.9	1232	4.3	1740	1.1
20 Sa	0038	4.3	0557	0.9	1310	4.3	1816	1.1
21 Su	0115	4.2	0633	0.9	1350	4.3	1853	1.1
22 M	0154	4.2	0712	0.9	1434	4.3	1934	1.2
23 Tu	0238	4.1	0756	0.9	1520	4.3	2021	1.3
24 W	0326	4.1	0844	1.0	1612	4.2	2114	1.4
25 Th	0421	4.0	0942	1.1	1709	4.2	2218	1.4
26 F	0524	4.0	1048	1.2	1813	4.2	2328	1.4
27 Sa	0634	4.0	1201	1.3	1918	4.3		
28 Su	0037	1.3	0743	4.2	1309	1.2	2019	4.5
29 M	0143	1.1	0848	4.4	1411	1.0	2117	4.6
30 Tu	0242	0.9	0948	4.6	1509	0.9	2213	4.7

JULY

Day	Time	m	Time	m	Time	m	Time	m
1 W	0338	0.7	1045	4.7	1602	0.8	2307	4.8
2 Th	0429	0.6	1139	4.7	1652	0.8	2357	4.7
3 F	0516	0.6	1230	4.7	1736	0.8		
4 Sa	0041	4.7	0559	0.7	1316	4.7	1820	1.0
5 Su	0123	4.6	0641	0.9	1400	4.6	1903	1.1
6 M	0205	4.5	0725	0.9	1446	4.6	1948	1.3
7 Tu	0247	4.3	0810	1.1	1532	4.5	2038	1.5
8 W	0336	4.2	0901	1.3	1621	4.4	2131	1.7
9 Th	0430	4.0	0958	1.6	1717	4.2	2234	1.8
10 F	0534	3.8	1103	1.7	1817	4.0	2342	1.8
11 Sa	0645	3.7	1212	1.8	1919	4.0		
12 Su	0048	1.7	0754	3.8	1315	1.7	2017	4.0
13 M	0146	1.6	0853	3.9	1409	1.6	2108	4.1
14 Tu	0235	1.5	0942	4.1	1454	1.5	2154	4.3
15 W	0317	1.3	1025	4.2	1534	1.3	2232	4.4
16 Th	0355	1.2	1102	4.3	1611	1.2	2310	4.4
17 F	0430	1.0	1137	4.3	1647	1.1	2345	4.4
18 Sa	0506	0.9	1212	4.4	1725	1.0		
19 Su	0021	4.4	0543	0.8	1251	4.4	1804	1.0
20 M	0059	4.4	0622	0.7	1334	4.5	1845	1.0
21 Tu	0140	4.4	0703	0.7	1420	4.6	1925	1.0
22 W	0224	4.4	0745	0.7	1506	4.5	2009	1.0
23 Th	0311	4.3	0831	0.8	1554	4.4	2055	1.1
24 F	0402	4.2	0922	1.0	1646	4.3	2151	1.3
25 Sa	0501	4.1	1024	1.2	1748	4.2	2259	1.4
26 Su	0612	4.0	1137	1.4	1853	4.2		
27 M	0015	1.4	0728	4.0	1252	1.5	2001	4.3
28 Tu	0130	1.3	0840	4.2	1402	1.3	2105	4.4
29 W	0234	1.1	0942	4.4	1501	1.1	2203	4.5
30 Th	0331	0.9	1039	4.5	1554	1.0	2255	4.6
31 F	0421	0.7	1130	4.6	1642	0.9	2344	4.6

AUGUST

Day	Time	m	Time	m	Time	m	Time	m
1 Sa	0506	0.7	1216	4.6	1725	0.9		
2 Su	0026	4.6	0547	0.7	1259	4.7	1807	0.9
3 M	0107	4.5	0625	0.7	1340	4.7	1846	1.0
4 Tu	0145	4.5	0703	0.9	1419	4.6	1925	1.2
5 W	0223	4.4	0742	1.0	1459	4.5	2005	1.3
6 Th	0304	4.3	0824	1.2	1539	4.4	2048	1.5
7 F	0348	4.1	0910	1.5	1622	4.2	2138	1.7
8 Sa	0440	3.9	1005	1.7	1715	4.0	2240	1.9
9 Su	0544	3.7	1113	1.9	1819	3.8	2353	1.9
10 M	0701	3.6	1228	2.0	1929	3.8		
11 Tu	0105	1.9	0815	3.7	1336	1.9	2033	3.9
12 W	0206	1.7	0914	3.9	1430	1.7	2127	4.1
13 Th	0254	1.4	1001	4.1	1514	1.4	2212	4.2
14 F	0333	1.2	1039	4.3	1552	1.2	2250	4.4
15 Sa	0411	1.1	1115	4.4	1630	1.2	2325	4.5
16 Su	0447	0.8	1151	4.5	1709	0.9		
17 M	0002	4.5	0527	0.7	1231	4.6	1749	0.8
18 Tu	0041	4.6	0609	0.6	1315	4.7	1830	0.8
19 W	0124	4.6	0651	0.6	1401	4.7	1912	0.7
20 Th	0209	4.6	0733	0.5	1447	4.7	1953	0.8
21 F	0256	4.5	0817	0.7	1531	4.6	2038	1.0
22 Sa	0346	4.3	0904	1.0	1623	4.4	2131	1.3
23 Su	0444	4.1	1004	1.4	1723	4.2	2240	1.5
24 M	0558	3.9	1121	1.6	1836	4.0		
25 Tu	0004	1.6	0724	3.9	1246	1.7	1954	4.1
26 W	0125	1.5	0835	4.1	1359	1.5	2102	4.2
27 Th	0232	1.2	0941	4.3	1459	1.3	2158	4.3
28 F	0326	1.0	1033	4.5	1548	1.1	2246	4.5
29 Sa	0411	0.8	1117	4.6	1631	0.9	2329	4.5
30 Su	0451	0.7	1157	4.7	1711	0.9		
31 M	0008	4.6	0530	0.7	1236	4.7	1749	0.9

Fig 67(b) Tide table Portsmouth.

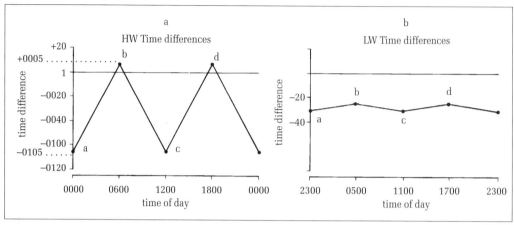

Fig 68 Full Graph. Time differences of Yarmouth and Portsmouth.

above, This time the graph is entered with the time of low water Portsmouth, and the time difference for low water Yarmouth is extracted. For example, the tide table for Portsmouth on 8 June gives the time of morning low water as 0837 GMT; entering Fig. 69 with this value gives a predicted time difference of −28 mins.

The small amount of time required to produce these two graphs is time well spent; however, in the LW case above, interpolating between 25 and 30 mins is just a bit over-zealous and has been done here merely to show the principle involved. In most cases a little mental arithmetic is all that is required to obtain

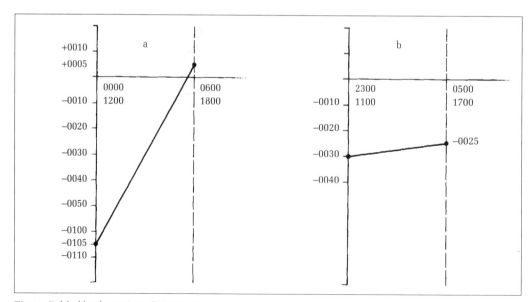

Fig 69 Folded back version of Figure 68.

times of high and low water for a secondary port.

Height Differences

The next four columns in the secondary port data deal with height differences:

At MHWS Portsmouth (4.7m), the HW at Yarmouth will be 1.6m lower (−1.6). At MHWN Portsmouth (3.8m), the HW Yarmouth will be 1.3m lower (−1.3). At MLWN Portsmouth (1.9m), the LW at Yarmouth will be 0.4m lower (−0.4). At MLWS Portsmouth (0.8m), the LW Yarmouth will be the same level (0.0).

The above values are used to produce the combined LW/HW graph of Fig. 70. From this graph the difference in height at Yarmouth can be extracted for any height of high water or low water at Portsmouth. For example, Fig. 67b predicted a Portsmouth HW level of 5.1m at 2345 GMT on the 4 May; extrapolating the HW graph of Fig. 70 predicts a difference at Yarmouth of −1.7m, so for 5.1m at Portsmouth, Yarmouth HW is 3.4m; quite a considerable difference.

Three final thoughts on secondary port difference:

(i) Plotting is not always necessary, and mental arithmetic can produce the interpolation required to find the difference.

(ii) The tide heights quoted in the almanac are mean spring and neap levels, therefore some tides will be higher/lower than those quoted. In such cases the graph line will need to be extended, or the mental effort suggested in (i) above will be one of extrapolation.

(iii) In some cases the spring and neap differences are nearly the same value, as in fact are those for low water Yarmouth, −25 and −30 mins; it would not be unreasonable to assume that the low water difference is −30 mins for all levels of low water.

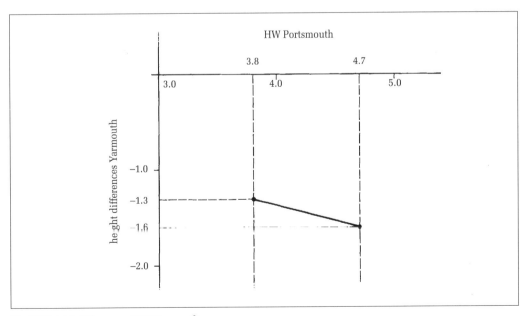

Fig 70 Height differences HW Yarmouth.

8

TIDAL STREAMS

In previous chapters we saw how the passage of the moon across the sky caused the tidal waters of the world to rise and fall as the earth rotated on its diurnal 24-hour cycle. In UK waters, semi-diurnal tides – that is, high waters occurring at intervals of around 12 hours and 50 minutes, separated by low water levels in between – give rise to a horizontal movement of water. This movement supports the vertical changes in height from HW to LW, and has two features:

1) The rate of movement generally referred to as the 'drift', or the distance a vessel is moved in one hour by the tidal flow.

2) The 'set', or the direction in which a tidal flow is moving. Along a coastline the set of the tide tends to follow the coast, changing its direction of flow every six hours or so. The horizontal flow of water with its set and drift is usually known as the tidal stream.

Typically, a UK tide rises for six hours as water floods towards the high-water level. It stands, or is said to be slack, at this level for a short period at high tide, then falls for six hours towards the low-water level, as the waters ebb away. There is a second stand, again for a short period, at low water before the next rise begins. Fig. 71 repeats the sinusoidal characteristic picture of a twelve-hour tide, and shows two 'stands', one at high and one at low water; the change in tidal height around these periods is almost zero, whereas at mid-tide periods the rate of vertical rise or fall of the tide is at a maximum. It follows, theoretically from this, that the horizontal movement of water must be almost zero at the times of high and low water, but will flow at the fastest rate at mid-tide periods.

The tidal streams are extremely important to the yachtsman simply because, if he can plan to make his passage along a coast at a time when the tidal streams are flowing in the direction in which he wishes to go, his speed over the ground (SOG) will be the sum of the tidal stream-rate and the yacht's speed through the water. Tidal streams around the UK are typically zero to 2 knots, and since your average – honest – yachtsman's speed through the water is 4 to 5 knots, a fair tide – that is, one going with you – will produce a best SOG of 6 to 7 knots; a foul tide, on the other hand, would result in a worst SOG of only 2 knots. The answer, of course, is to 'work your tides' wherever possible to ensure a swift passage.

In some areas of the UK, tidal stream rates are very much greater than the

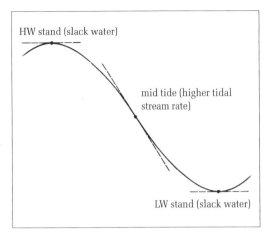

HW stand (slack water)

mid tide (higher tidal stream rate)

LW stand (slack water)

Fig 71 Twelfths rule tide.

typical 2-knot figure quoted above. This is particularly true at equinoxial spring tides, when very fast, fair tidal streams, that are indeed very fair to the planning of a fast passage, occur. In cases of long passages when periods of a foul tide cannot be avoided it may be advantageous to enter a convenient mid-passage port, or to anchor to await the turn of the tide.

The prudent skipper will be well aware that his SOG during a fast, fair tidal stream needs to be carefully considered lest the stream causes him to watch helplessly, despite all that sails and engine can do, as his planned destination slips rapidly away as the stream carries him further and further down-tide. Far better to have the mid-tide rate assisting you on the journey so that you arrive at your destination at or near a stand of the tide to facilitate an easy entry to the desired port. In cases where an approach at mid-tide has to be made, a transit based on the entrance and some background object will prove most useful, ensuring that your course over the ground is going to put you at the entrance, even though your heading may lead you to think otherwise.

The hourly direction and speed of the tidal stream – the set and drift – are published as a series of chartlets in Admiralty tidal stream atlases, and the yachtsman needs to obtain the atlas covering the particular area of interest. For example, tidal atlases are produced for many separate areas, the English Channel and the Solent being just two of them. The tidal stream data is also shown on charts as a table of 'set and drift' values.

The Admiralty Tidal Stream Atlas

Fig. 72 shows an extract from a tidal stream atlas covering the sea area Littlehampton to Burnham-on-Crouch. This particular atlas is based on the times of high water Dover. Tidal atlases covering other areas may well be based on a different standard port.

To use the information shown on Fig. 72 the following operations are performed:
(1) The time of HW Dover, which is closest to the time period of interest, is extracted from the relevant tide tables and copied, lightly in pencil, alongside the HW Dover chartlet. Fig. 72 chartlets cover only the hours before HW Dover, but the full atlas has a complete page for each chartlet and would also cover the hourly intervals after HW Dover; on each chartlet the relevant ship's time would be lightly printed.
(2) The direction arrows would be examined; their length, thickness and density indicate the relative strengths of the stream, and the arrowhead the direction of the 'set' (the direction in which the tidal stream is running). The two two-digit numbers alongside most arrows give

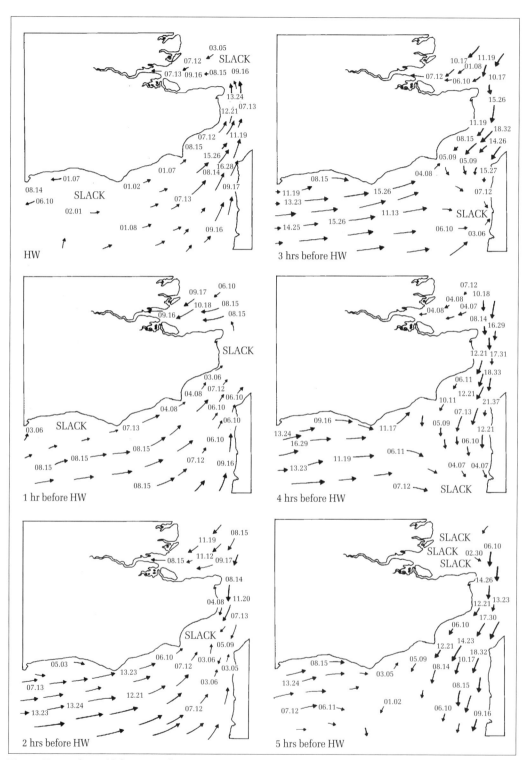

Fig 72 Extract from tidal stream atlas.

the 'drift' (the speed of the stream at that place and at that time). The two digits either side of the separating comma represent the mean neap and spring drift. For example, the arrow and associated figures for the stream off Folkestone at HW Dover shown as 07,12, indicate that the stream is running along the coast in a NEly direction, the predicted drift at a mean neap tide is 0.7 knots and at mean springs is 1.2 knots (the decimal point is simply not shown on the chartlets). In practice it is convenient to assume that the arrow's data holds true for one full hour, from 30 mins before the hour stated to 30 mins after the hour stated.

Examine Fig.72 again, and the chartlet for the time of two hours before HW Dover; the stream off Dover is marked as 'slack', indicating that the stream is about to change direction; before this time the stream is shown going southerly, afterwards it is shown to be northerly. A skipper wanting to sail north from Dover to Ramsgate, a passage of some 15 nautical miles, on a day when the wind forecast will give the yacht an assumed 5 knots through the water, would do well to take advantage of a favourable tide by leaving at, say, one hour after HW, because, as can be seen from Fig. 72, the tide is then fair for the journey. Assume that on the day in question a spring tide HW at Dover is predicted for 1100 BST: the chartlet for one hour after HW, covering the period 1130 to 1230 BST, is fair at 1.3 knots; the next hour is also shown fair at 2.3 knots, and the next is still fair, but only at 1.4 knots.

By leaving his berth at 1100 the skipper should be outside the harbour at 1130 to pick up the favourable stream. After two hours the yacht will have travelled 5.0+5.0+1.3+2.3 = 13.6 nautical miles over the ground, and in the next twelve

minutes its speed and the still favourable stream should see it off Ramsgate. Without the assistance of the tide the passage time would be three hours, and in a foul tide it could well be four hours.

Charted Tidal Streams

All charts include a table of tidal stream set and drift for the area of that chart. The tidal stream table extracted from chart 5061, Dover to North Foreland, is shown at Fig. 73. The standard port is Dover. In use, the left-hand column of the table is entered with the time of HW Dover, or with the appropriate hour before or after HW Dover. Note how the columns of the table are headed with a 'tidal diamond', each diamond having a capital letter in its centre and an associated latitude and longitude position. The lat. long. position printed alongside a tidal diamond pinpoints a position on the chart where the set and drift figures in that particular column of the table apply. The lettered tidal diamond is also shown again at the actual position on the chart. The first column, for example, is tidal diamond A, and its position is 51°23'9N 1°20.5E; entering this column at the HW point gives the set as 250°(T), the spring tide drift 1.7 knots and the neap tide drift 1.1 knots. Therefore at position 51°23'9N 1°20.5E on the chart, the set and drift, at the time of HW Dover, is 250°(T) at 1.7 knots for a spring tide, and 250°(T) at 1.1 knots for a neap tide.

The set and drift figures in the table are assumed to hold good for one hour: that is, from 30 mins before the stated time to 30 mins after the stated time; figures given for HW Dover hold for HW–30 mins to HW+30 mins.

The table of Fig. 73 lists fourteen tidal diamonds, each one representing a position on the chart. A navigator working up a course to steer, or an E.P. (estimated position) in the vicinity of a tidal diamond, simply extracts the data from the quoted diamond at the relevant time. In practice, of course, a yacht will spend most of its time sailing between several tidal diamonds, and the yachtsman will need to the interpolate data from the appropriate diamonds.

Here is an example: at a time of springs, a yacht sailing midway between tidal diamond A and B at a time of Dover HW−3, would extract the following from the table of Fig. 73:

Diamond A: set 250° drift 1.2 knots
Diamond B: set 213° drift 2.8 knots

Interpolation gives:

set (250-213°)/2+213 = 37°/2+213 = 18.5+213 = 231°

drift (2.8-1.2)/2+1.2 = 1.6/2+1.2 = 2 knots.

The stream data obtained from tidal chartlets and chart tables quoted above are for spring and neap tides; to obtain stream figures for periods outside these tides, the yachtsman must either do further interpolations or use a 'computation of rates' table, such as the one shown at Fig. 74, where the table is based on HW Dover. The vertical scales are marked off in 'range of tide', and the horizontal scales in 'tidal stream rates' or drift. Note also that the range of mean neap and mean spring tides for Dover are highlighted by dotted lines across the table,

Tidal Streams referred to HW at DOVER

Hours	Geographical Position	(A) 51°23'·9N 1 20·5E		(B) 51°20'·3N 1 34·3E		(C) 51°20'·1N 1 30·0E		(D) 51°19'·7N 1 27·7E		(E) 51°18'·2N 1 34·4E		(F) 51°18'·2N 1 31·7E		
Before High Water 6	064	1·0	0·6	199	2·0	1·2	191	1·0	0·6	203	1·2	0·7	168	1·1 0·6	218	1·1 0·8	−6
5	070	0·5	0·3	204	2·6	1·5	200	1·5	0·9	203	1·3	0·7	182	1·5 0·9	222	1·7 1·0	−5
4	239	0·5	0·3	208	3·1	1·7	202	2·0	1·1	210	1·7	1·0	188	2·0 1·1	228	2·3 1·3	−4
3	250	1·2	0·7	213	2·8	1·5	199	2·2	1·3	208	1·9	1·1	192	2·1 1·2	229	2·1 1·2	−3
2	250	1·5	0·9	222	1·5	0·8	203	1·7	0·9	215	1·4	0·8	215	0·8 0·4	230	1·2 0·7	−2
High Water	250	1·7	1·1	357	0·8	0·5	340	0·6	0·3	005	0·6	0·4	350	1·2 0·7	016	0·6 0·3	−1
	250	1·7	1·1	015	2·5	1·4	007	2·4	1·3	021	2·2	1·2	002	2·4 1·4	039	2·0 1·1	0
After High Water 1	255	0·9	0·8	023	3·2	1·8	015	2·4	1·3	030	2·3	1·3	008	2·1 1·2	044	2·3 1·3	+1
2	070	0·5	0·3	029	2·9	1·6	023	2·0	1·1	032	1·9	1·1	012	1·4 0·8	045	2·1 1·2	+2
3	070	1·8	1·1	044	2·2	1·3	031	1·4	0·8	043	1·2	0·7	010	0·8 0·4	048	1·6 0·9	+3
4	070	1·9	1·2	059	1·2	0·7	051	0·7	0·4	073	0·4	0·2		0·0 0·0	057	0·8 0·5	+4
5	066	1·5	0·9		0·0	0·0	135	0·4	0·2	195	0·6	0·3	142	0·5 0·3	168	0·4 0·2	+5
6	085	1·1	0·7	197	1·4	0·8	195	0·8	0·5	203	1·1	0·6	183	0·9 0·5	210	1·0 0·5	+6

	(G) 51°17'·9N 1 29·4E		(H) 51°16'·3N 1 27·4E		(J) 51°15'·2N 1 32·6E		(K) 51°13'·3N 1 26·6E		(L) 51°13'·0N 1 36·4E		(M) 51°10'·5N 1 32·2E		(N) 51°09'·0N 1 27·8E		(P) 51°06'·6N 1 20·4E		
−6	206	1·5 0·8	195	2·0 1·1	223	1·9 1·1	181	1·6 0·9	190	0·9 0·5	225	1·7 1·0	212	2·2 1·2	224	2·3 1·3	−6
−5	214	2·1 1·2	197	2·6 1·5	228	2·5 1·4	183	2·1 1·1	191	2·3 1·3	225	2·5 1·4	213	2·2 1·2	231	2·5 1·4	−5
−4	218	2·5 1·4	197	2·8 1·5	226	3·1 1·7	186	2·1 1·2	195	3·1 1·7	222	3·1 1·7	216	1·8 1·1	233	2·4 1·3	−4
−3	217	2·5 1·4	202	2·4 1·3	225	3·1 1·7	188	1·9 1·0	196	3·2 1·8	219	2·9 1·6	228	1·3 0·8	225	1·5 0·8	−3
−2	219	1·5 0·9	215	1·0 0·6	231	1·2 0·7	190	0·8 0·5	195	2·0 1·1	224	1·4 0·8		0·0 0·0	075	0·3 0·2	−2
−1	008	0·7 0·4	012	1·3 0·7	040	1·3 0·7	007	0·9 0·5		0·0 0·0	014	0·5 0·3	032	1·2 0·7	058	2·3 1·3	−1
0	024	2·3 1·3	017	2·7 1·5	041	2·7 1·5	001	2·1 1·2	013	1·3 0·7	040	2·2 1·2	038	2·0 1·2	083	3·9 2·2	0
+1	029	2·8 1·6	027	3·2 1·7	043	2·8 1·6	001	2·3 1·3	015	2·4 1·4	042	3·0 1·7	039	2·3 1·3	064	4·1 2·3	+1
+2	032	2·5 1·4	018	2·6 1·4	046	2·5 1·4	003	2·3 1·3	014	3·1 1·7	037	2·8 1·6	034	2·2 1·2	066	3·5 1·9	+2
+3	039	1·7 1·0	022	1·7 0·9	049	1·7 1·0	002	1·4 0·8	017	2·6 1·5	047	1·9 1·1	031	1·5 0·8	072	2·6 1·4	+3
+4	050	0·9 0·5	037	0·6 0·3	055	0·7 0·4	017	0·6 0·3	018	1·7 1·0	061	0·9 0·5		0·0 0·0	087	1·3 0·7	+4
+5	156	0·3 0·2	205	0·4 0·2	171	0·2 0·1	161	0·3 0·2	018	0·6 0·3	167	0·3 0·2	203	1·0 0·6	208	1·1 0·6	+5
+6	202	1·1 0·6	197	1·6 0·9	219	1·5 0·9	180	1·3 0·7	189	0·5 0·3	218	1·3 0·7	210	1·8 1·0	220	2·2 1·2	+6

Fig 73 Tidal diamonds.

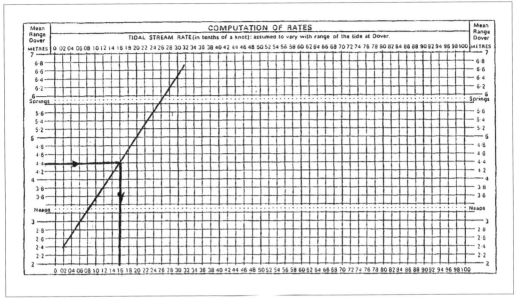

Fig 74 Computation of tidal stream rates.

mean spring and neap ranges being 5.9 and 3.3m respectively.

To use the compilation table, the tidal chartlets or tidal table is first used to obtain the neap and spring tides drift data, say 0.8 and 2.6 knots respectively. These values are plotted against the neap and spring ranges as shown at Fig. 80, then joined by a straight line. The navigator having calculated the range of the tide for the day at the charted position of interest, enters this value – say, 4.4m – into the vertical scale of the compilation table, and, from where this entry intercepts the drawn line, reads off the tidal drift figure; in this case Fig. 74 yields a rate of 1.6 knots.

When using the compilation table it should be remembered that the neap and spring ranges, shown dotted across it, are calculated from mean values of these tides, and since some spring tide ranges are greater and some neap tide ranges smaller than the mean values, the navigator may well experience a tidal range outside the mean values, in which case the intercept on the drawn line will be an extrapolation and not an interpolation.

In completing the above calculations, the set of the tide is assumed to be the direction quoted for neap and spring values in the table referenced.

9

CHARTWORK 2

In Chartwork 1, a yacht's position was plotted using only the compass course steered and the logged distance run; however, the resultant dead reckoning position (DR) is very rarely an accurate one because it does not take into account the sideways movement of the yacht through the water – leeway – nor does it allow for tidal set and drift.

Leeway is due to the fact that the wind entering the sails of a yacht not only pushes her forwards, but also tends to push her downwind at right-angles to her heading. The amount of leeway is depen-dent upon wind strength, point of sail and the underwater design of the ship's hull; it can be estimated by looking back over the stern of the boat: if leeway is present the yacht's wake will not be in line with her heading, but angled off to wind-ward; she is actually travelling over the ground on the 'wake course' and not towards her heading.

Leeway can be measured by taking a hand-bearing compass bearing of the yacht's wake, and subtracting this value from, or subtracting from it, the reciprocal of the ship's compass course (see Fig. 75).

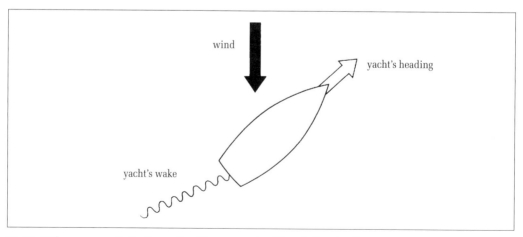

Fig 75 Wake course, entirely due to the sideways push of the wind. When beating into the wind the sideways push is at maximum.

Please note that the deviation of a hand-bearing compass is assumed to be zero. Also note that the yacht's wake course is the reciprocal of the wake's bearing taken over the stern. Leeway is at a maximum when beating hard to windward, and it falls to zero on a full downwind run; it is typically quoted as 10 degrees but will vary, not only for different points of sail but also from yacht to yacht, and the prudent skipper will measure its value whenever he can. The effect of leeway upon a yacht's movement over the ground can also be observed by looking ahead, that is in the direction of the yacht's heading, and preferably to a land feature or buoy. If the yacht appears to be crabbing across the feature or away from the buoy, then leeway may be the cause of that crabbing.

Estimated Position (EP)

A yacht is steered by her compass, and her compass course is the ship's heading that is, the direction in which she is pointing. The effect of leeway is to push the boat downwind such that her course over the ground, now called her 'wake course', is the compass course plus or minus the leeway. (The effect of leeway on a yacht's movement over the ground is shown at Fig. 76.)

The tidal set and drift will also affect a yacht's course over the ground. Where the tidal set is in the same direction as the yacht's course, her speed over the ground is the sum of her speed through the water and the tidal drift. Where the set is in the opposite direction, her speed over the ground is equal to her speed through the water minus the tidal drift.

In general, the tidal stream set will be at some angle, rather than dead ahead or astern to a yacht's wake course, and its effect then is best seen by constructing a diagram, technically called a vector diagram (see Fig. 77). It is constructed by first drawing the DR from the last position – line ab of Fig. 77 – then adding the wake course, allowing for leeway: it will have the same length as the ship's DR line, and is shown as line ac. Finally a line, technically called the tidal vector and which

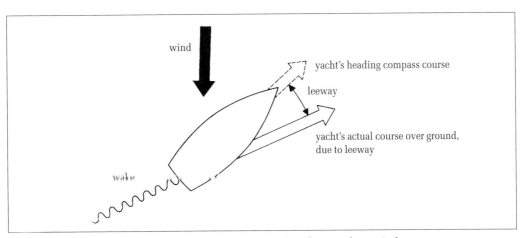

Figure 76 Leeway – The effect of leeway is to push the yacht sideways, downwind.

represents the tidal set and drift, is appended to the end of the wake course, line cd. The diagram is now complete, and has taken into account the following:

the compass course steered;

the logged distance run, through the water;

leeway, ie sideways push of the wind;

tidal stream, set and drift.

The charted position at the end of the tidal vector is called the 'estimated position' (EP) of the yacht, and, in the absence of a fix, 'working up' an EP is the most accurate method of plotting a yacht's position without the aid of astro-navigation or electronic plotting, as we shall see later. In Fig. 77 joining the points a and d produces the line ad, the actual course travelled by the yacht, called the ground course.

An EP should be plotted onto the chart at hourly intervals. However, although it is the 'best effort' that the navigator can make in determining his position because it takes all things into account, it is still only an estimated position, and questions such as: 'How well has the course been steered?' or 'How good was our estimation of leeway?' and 'How accurate was the interpretation of the tidal set and drift for our particular course and position?' will be constantly on the skipper's mind.

Verification of the yacht's position is a permanent quest, and again there will be constant questions such as: Does the echo sounder reading agree with the sounding on the chart at the EP? Are there any conspicuous shore features visible which also appear on the chart? Are there any buoys visible? Ideally everything visible and readable must confirm the EP and if it doesn't, then the EP is suspect and extra caution is required until the boat's position can be estimated more accurately.

One extremely accurate way of plotting a yacht's position is by taking bearings of conspicuous, charted shore features and transferring these bearings onto the chart. Ideally three shore objects are required, although two will do, and, as we shall see, when necessary a fix can be obtained using only one shore feature.

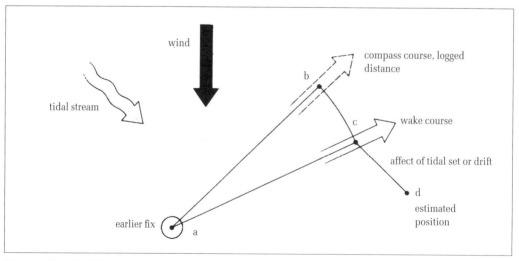

Fig 77   Effect of Tidal stream.

Three-Point Fix

The bearings of three conspicuous charted features are taken – for example, the church spire, a water tower and a flagstaff – and the reciprocal bearing of each one is plotted on the chart. If the bearings are taken carefully and accurately, the three lines will all cross at a point, that point being the position of the yacht from which the bearings were taken.

In practice, accurate bearings are difficult to take from the deck of a sailing yacht, and the lines will most likely cross over a small area producing a 'cocked hat', rather than at a single point. The yacht is assumed to be in the centre of the cocked hat, as shown at Fig. 78. Two bearings are all that are strictly necessary; however, the crossing of all three at one point, or over the small cocked hat area, gives confirmation that all three bearings are good.

Ideally the three objects should be about 30 to 50 degrees apart to give a good cut on the chart. They should not be more

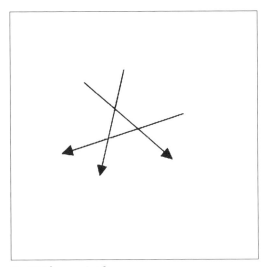

Fig 78 Three-point fix.

than 3 to 5 nautical miles away, since just a small error in the reading of the bearing of a distant object will produce a large cocked hat. The bearing of an object on the beam should be taken last, because this one will be altering at the fastest rate. Finally the objects of interest should be well scrutinized first; a crew-mate should jot down their identity, copying the bearings against them as you call them out in an agreed order.

Running Fix

This is sometimes called 'doubling the angle on the bow', and is a method which requires only one fixed conspicuous shore object. Before we describe how the fix is made, a little geometry theory will help to explain its validity: in any triangle the sum of the three included angles (angles A,B and C of Fig. 79a) is 180°. In any triangle where two included angles have the same value (angles A and B of Fig. 79b), the sides opposite those equal angles have the same length. In order to simplify the following construction we will assume that both leeway and tidal effects are zero.

To construct a running fix on the chart, a bearing is taken of the shore object using the hand-bearing compass, and the distance log reading is noted. The reciprocal bearing is plotted on the chart from the shore object such that it cuts the course line, and the resultant 'angle on the bow' is calculated. Fig. 80 shows a yacht on course 070° magnetic; the bearing of the shore object was measured as 040°M using the hand-bearing compass, giving an angle on the bow of 30° (70–40). The passage is continued until the angle on the bow has doubled, in this case

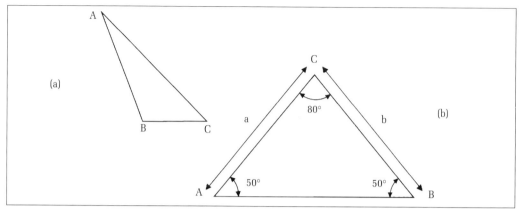

Fig 79 Angles of a triangle.

becoming 60°. Fig. 80 shows that this will be true when the object's bearing is 010° magnetic. When the object bears 010°M the distance log is again noted.

Fig. 80 is used to explain how the distance travelled by the yacht between the time of the two bearings is equal to her distance off the shore object at the time of the second bearing. If the logged distance was 5 n. miles, then at the time of the second bearing the position of the yacht will have been fixed at a bearing of 190°(M) from the shore object (reciprocal bearing of 010°(M)), distance 5 nautical miles.

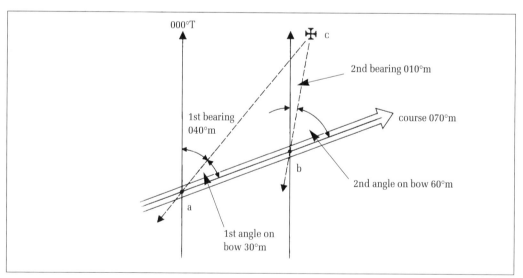

Fig 80 A running fix, at position a the church's magnetic bearing is 040°M, angle on the bow is therefore 070°–040°=30°. Point b is the position at which the angle on the bow has doubled to 60°, ie the object must have a magnetic bearing of 010°M. Point b is reached therefore when the observer notes a hand bearing compass readout of 010°M.

Transfer Position Line

This is a second method of fixing a yacht's position with only one shore object in view, and in this case we will assume that both leeway and tidal effects are present. Let us take a situation where the time is 1200 BST, the yacht's course is 070°(M), the object's bearing is 040°(M) and the log reading is noted as being 168 nautical miles. Fig. 81a shows the situation; the wind is from the north, and leeway is 10°. At 1300 BST a second bearing is taken, of the same shore object and noted as 350°(M), the log being 173.5 nautical miles. From 1200 to 1300 BST the tidal set and drift was 135°(T) at 1.6 knots. What is the yacht's position at 1300 BST?

The reciprocal of the first bearing is laid off on the chart from the shore object. Assuming that the yacht's passage has so far been in extreme weather conditions and that her plotted position is very suspect, the skipper at least now knows with certainty that she is somewhere on the bearing line at the time it was taken, and it may even be possible to estimate her distance from the shore with the naked eye. Howsoever a point is chosen on the plotted bearing line as the yacht's 1200 hour position, the EP for the next hour is plotted from that point. At 1300 the second bearing is taken and plotted.

Fig. 81a shows a possible and likely 1300 situation. The problem is that the EP should lie on the second bearing line – after all, the two bearings were carefully and accurately taken, and the EP very carefully plotted – but it doesn't. The problem lies in the skipper's guesstimation' of the 1200-hour point on the first bearing line, that can now be seen to be an error. The solution is to draw a new

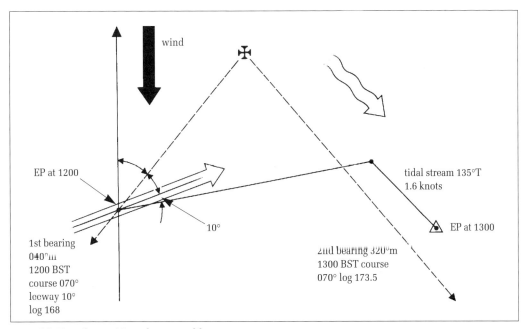

Fig 81(a) Transfer position plot – a problem.

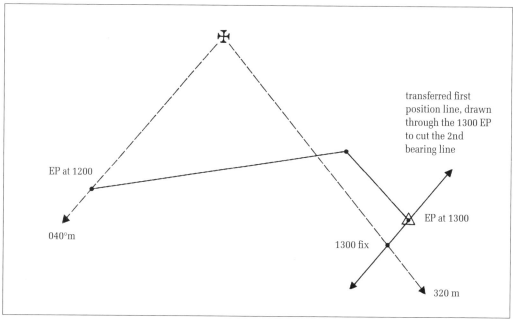

transferred first
position line, drawn
through the 1300 EP
to cut the 2nd
bearing line

EP at 1200

040°m

EP at 1300

1300 fix

320 m

Fig 81(b) Transfer position plot – a solution.

line, parallel to the first bearing line, which passes through the 1300 EP and cuts the second bearing line. In other words, 'transfer the first position line' so that it passes through the 1300 EP. The cut on the second bearing line is (it must be) the 1300-hour position of the yacht.

Fig. 81b – an uncluttered copy of Fig. 81a – shows this new line, which in reality is the transferred first position line. It also shows the correct 1300-hour position

of the yacht – in fact, this is a fix. This figure could also be used to show that if the 1200 to 1300 ground track were drawn in reverse order – that is, from the known accurate 1300-hour position – it would cut the first bearing at what must have been the yacht's 1200-hour position. Thus honour and accuracy would be satisfied: a closed triangle formed by the two bearing position lines and the yacht's plotted ground track.

10

PILOTAGE

Pilotage concerns the passage of a yacht from its port of departure along the coast to its port of arrival. Amongst the questions to be asked are these:

(1) Is the predicted weather suitable for the yacht, the crew and the experience of the skipper?

(2) Are suitable charts inboard?

(3) Is it a simple, single course passage?

(4) Are suitable nautical almanacs and pilot books on board?

(5) Can the distance involved normally be completed in one daylight run? Or is a night passage involved?

If all the answers are affirmative, the passage plan outline can be drawn up, considering the following points:

(1) Are the departure and destination ports open at all states of the tide?

(2) Is there sufficient water at all states of the current tides? If not, is there a suitable waiting pontoon or sheltered anchorage in the vicinity?

If all the answers are affirmative, divide the distance to run by the estimated speed to obtain the time of the passage, and then choose your time of departure.

Detailed Planning

From the small-scale charts of the two ports involved, examine the soundings; if they exceed the yacht's draught at all positions of interest, no further departure/arrival port tidal calculations are necessary. If, however, the draught plus safety depth exceed the soundings, then tidal curves must be completed to find the departure/entry 'time window' for the port involved. If either port has a locked entry, then reference to the pilot book will give times of entry/departure ±HW.

If it is the destination port that has the time limitation, we must subtract the estimated time of the passage from this window to obtain our earliest/latest times of departure. If both ports have a time limitation and our predicted passage time does not fit in, then unfortunately we may have to 'wait out' the time difference.

Example

The tidal curve for a port of destination is shown at Fig. 82. A yacht drawing 2.1m, and a skipper who believes that 0.5m is a good safety margin, wish to enter this port after completing a 25-mile passage at an estimated speed of 5 knots. Assume

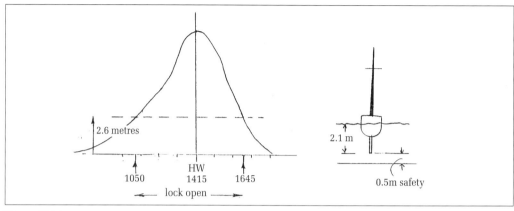

Fig 82 Tidal curve.

that the port of departure is locked, that HW occurs at 0805, and the lock is open at HW±0140.

From Fig. 82 we can work out that the destination port can be entered by this yacht at any time between 1050 and 1645. However, it would be good seamanship to leave a time margin of safety at the falling tide end of this window and restrict our time of entry to between 1050 and, say, 1600. Using the earliest time of entry, 1050, and subtracting our expected 5 hours of passage, we would need to leave at 0550. However, the lock at this port would only be open between 0805±0140, ie between 0625 and 0945. If we leave as soon as the lock opens and travel for our 5 hours, we would arrive at 1125, well within the destination port's favourable time.

Coping with Problems

Obviously problems can arise to spoil our nicely laid plans: the wind may die, it may be right on the nose, or it may be so strong that we have to run for shelter. Always be sure to have on board the relevant small scale charts for all the interme-diate ports on the particular passage; make sure that the engine is kept well maintained so that, should we require to use it, it will work first time; and check that there is plenty of fuel. The course line on the chart will indicate the availability of navigation aids such as charted transits, and those that we can define ourselves – a distinct and conspicuous shore feature in line with an easily identifiable inland mark such as a church spire or water tank, measured mile markers, and even normal port and starboard hand-markers – all these will help to confirm our position.

Charted soundings with allowances for height of tide may be used in conjunction with a transit to transform a transit bearing to a transit/depth fix. In some cases an isolated sounding of an undersea hill or valley can be useful in confirming or otherwise our estimated position.

Planning Longer Passages

The above exercise was relatively trouble free and the planning was straightforward, the passage representing a short, pleasant day's sail between two easily accessible ports. Where circumstances are more complex – a much longer journey of, say, 135 n. miles at an estimated speed of 5 knots, that is 27 hours of sailing time – more detailed planning of the passage is required, and many more questions have to be answered before departure. Even for distances greater than 40 n. miles, night sailing will be involved at some stage and it would be desirable, though not essential, to arrive at the destination in the early hours; the navigation lights as we approach will identify the destination port positively, and, as the dawn breaks, reference to the relevant pilot book will identify any conspicuous shore features.

The skipper will no doubt want to be awake as the destination is approached, and this presupposes that he will have slept during the night. Watch keepers are therefore required, and must have a thorough knowledge of the IRPCS, especially ships' lights, day shapes and buoyage and so on. The first so that they avoid all those things that they have to avoid, like fishing vessels; the second and third, that they may pass on the safe side of all the buoyage marks that will be encountered on the journey. Watch keepers must also plot the estimated position of the yacht every hour, and at every course change. They must also have the confidence to alert their sleeping skipper whenever his greater experience is required.

The passage may be long enough for the weather to turn nasty, even though the long-term forecast is favourable as you leave. So do you have sufficient foul weather kit, life jackets and safety harnesses for everyone?

Is the wind or sea state likely to increase during the dark hours? If so, the night 'sail plan' needs to be considered: unless you are in a hurry to make your destination, you might consider reducing your sail area before dark, especially if your crew are not experienced in reefing the main or changing the jib

What about sea-sick pills? Not everyone is prone to the malady, but the crew need to know how to recognize the early symptoms of *mal de mer* within themselves, and the skipper must reduce the possibility of such illness by ensuring that all are warm, well fed – yes well fed – before the problem arises; also that they are all kept occupied.

Forty hours of sailing means several tide changes, and this means that the tidal streams will sometimes be against you. A foul tidal stream represents those occasions when a 30ft yacht, which may well be doing 5 knots through the water, is doing very little speed over the ground. Such circumstances may well prevail if the yacht is attempting to pass a headland at the time of a foul tide. This would probably be a good time to anchor for a pleasant meal in the cockpit. If you can plan so that the passing of the headland is associated with a fair tide, then you are indeed very fortunate and you will pass the point at the best speed.

Overfalls

Your chart will show the presence of any overfalls at headlands, and the pilot book will inform you of the extent of the danger posed by them (if any), and how far you may have to stand out to seaward to avoid their effect. Overfalls can be quite severe; they usually occur where, over a considerable time period, sea erosion has caused the collapse of the original much more extended headland, so that the sea bed now consists of the residue of the original land, which may rise steeply from the sea bed itself. This steepness effectively provides a sudden wall to the lower waters of a tidal stream, causing them to rise to the surface; in places where they are sufficiently strong, the waters become so confused that the surface may seem literally to boil. Usually these overfalls begin and end very abruptly, and extend in a straight line seaward from the remaining headland.

Please note that overfalls can be very dangerous to yachts: the turbulent seas will throw the boat around with great force, and the helm may well have but little control; the main could easily be backed so that the boom comes across the cockpit in an unwanted gybe, and, down below, crew and unsecured items will be thrown around the cabin risking bruising and fractured bones. In short, treat overfalls with great respect. The pilot will also indicate the presence of an inshore passage.

Other Navigation Aids

Studying the chart before you depart will acquaint you with other navigation aids to assist you in a safe passage, including charted transits and unique soundings – undersea hills and valleys standing alone in an area of a generally flat sea bed. The chart will also highlight traffic separations schemes (TSS), and you must stay out of these specified channels unless you need to cross them. If you do need to cross a TSS, you must do so at right-angles to the general direction of the traffic using the TSS – that is to say, your heading must at 90 degrees to the direction of vessels using the TSS. Your actual course over the ground is immaterial, because the intention of this most important rule is that you offer the biggest visual and radar image to vessels within the TSS.

11

SAFETY

Safety at sea – of the crew, of the yacht, of third party persons and of equipment – is of paramount importance; moreover it is the sole responsibility of the skipper and cannot in any way be delegated. The question arises, however, as to what happens when the skipper is down below at the chart table, for instance, or if he is quite legitimately asleep? The answer is always the same:

> Safety is the sole responsibility
> of the skipper.

It follows therefore that the skipper must ensure that all persons on his yacht receive sufficient training and practice in all aspects of safe personal behaviour, and in the proper and safe operation of any equipment that they may use on the yacht. Whilst on deck he is obviously in charge of all deck work, but he also remains responsible for the safety of the people below deck, and for the correct and safe operation of all equipment below.

When a skipper brings new crew on board his yacht he should give all members a safety briefing, as we shall discuss later. The briefing will cover many items, but one of the most important concerns

the gas oven, essential for the jollification of the crew, and used by all members – after all, it is a common enough piece of shore-side domestic kitchen equipment. The gas oven at sea, however, needs to be seen as a source of potential danger, the problem being that gas is heavier than air so that any leakage will drop to the bilges and accumulate in the yacht, with disastrous consequences should anyone produce a naked flame. The skipper must therefore have trained every member of the crew in its safe operation before it is used. Procedure would be as follows:

The gas bottle(s) will be stored external to the cabin and its (their) storage compartment will vent outside the yacht. When arriving at the yacht, the very first operation should be to pump out the bilges in case there has been a leakage of gas during the previous time period. Should there be any water in the bilges this will be expelled first. When water is not present or when all the water has been expelled, a further ten to twenty pumps will ensure that any accumulated gas is also expelled.

The gas bottles themselves will each be fitted with a 'turn-off' valve and there is usually a second turn-off valve sited close to the cooker, the rules controlling their use represent the 'gas routine of the yacht'. The skipper will explain this and the reasons for its importance and every skipper will have his own routine. Either both turn-offs will normally be off, or perhaps only the one next to the cooker after it has been used, but the essential order is that he/she who turns it on *must* turn it off immediately after use; anyone finding the valve on when the cooker is not in use should make a fuss, not in order to embarrass anyone but so that the gas routine is brought back to everyone's mind.

The bilges should be pumped clear each morning, and when leaving the yacht for lengthy periods all gas turn-offs should be closed. Many yachtsmen find a gas 'detector and isolator' a very comforting piece of safety equipment.

Safety Equipment

Before leaving the berth, a skipper's safety briefing should include the instructions and advice itemized below; it is not an exhaustive list of safety features and fittings, but may well serve as a guide to new skippers.

Fire-Fighting Equipment

The location of the yacht's fire-fighting equipment should be pointed out, and its operation explained. It is no bad thing for the crew members actually to read the user instructions of all the fire extinguishers fitted. The siting and use of a fire blanket needs to be explained, and make sure that fire blankets are not positioned where the potential user would need to stretch across, say, the cooker to get them.

The Loo

Not essentially a piece of safety equipment, but certainly one whose operation may be a mystery to the new sailor. Unblocking a blocked loo at sea, particularly one that has been pressurized by misuse, is not a task anybody wants, and can only be avoided by instruction if not by demonstration. Once the operation has been explained, the facts are that, apart from proper loo paper, only those things which you have first eaten are allowed into the loo bowl.

VHF Radio

The very high frequency (VHF) radio is used to communicate with family and friends through shore radio stations, with marinas to request an overnight berth, and with the coastguard for small boat safety traffic. A user must be under the supervision of a licensed operator. More importantly, VHF is used to call for help and assistance. VHF operations follow special procedures, particularly when help is being requested.

THE DISTRESS SIGNAL 'MAYDAY'

The most important use of the VHF radio is to transmit the internationally recognized distress signal 'MAYDAY', and this has absolute priority over all other transmissions. It is made on Channel 16 by the vessel in distress, and on hearing a mayday call all other stations must immediately stop transmitting on this channel but should continue to listen on it.

The mayday signal is made only when there is *grave* and *imminent* danger to the *vessel* or a *person* on board, when *immediate* assistance is needed. The mayday signal has two parts:

(1) The MAYDAY call:

Switch on the radio, select Channel 16, turn to high power, press the 'push to transmit button'.

Speak as clearly as you can the following;

> MAYDAY MAYDAY MAYDAY
> This is, YACHT'S NAME three
> times.

(2) The MAYDAY message:

This should be transmitted, after taking a deep breath, immediately following the mayday call:

> MAYDAY
> YACHT'S NAME
> POSITION
> NATURE of DISTRESS and type of
> ASSISTANCE required
> Other RELEVANT information.

EXAMPLE

> MAYDAY MAYDAY MAYDAY
> THIS IS FOXGLOVE FOXGLOVE
> FOXGLOVE
> (deep breath)
> MAYDAY
> FOXGLOVE
> 56 DEGREES 20 MINUTES NORTH
> 05 DEGREES 10 MINUTES WEST
> HOLED AND ABANDONING TO
> LIFERAFT
> FOUR ADULTS
> OVER.

In describing the above procedure we said 'switch on the radio'; however, normally the VHF radio would already be on when a yacht is at sea, and it would, where fitted, be on 'dual watch' – that is, listening on a user-selected channel but automatically going over to 16 when traffic is received on that channel. If dual watch is not fitted, then a continuous VHF radio watch should be kept on channel 16.

When a yachtman hears/receives a distress call, the associated distress message should be written down. No immediate action, other than a listening watch, should be taken. However, if after two or three minutes there has been no acknowledgement from a 'shore radio station' or another vessel, then the vessel hearing the distress should make an acknowledgement in the following terms:

> MAYDAY
> NAME of vessel in DISTRESS three
> times
> THIS IS NAME of vessel
> ACKNOWLEDGING three times
> RECEIVED MAYDAY.

EXAMPLE

> MAYDAY
> FOXGLOVE FOXGLOVE
> FOXGLOVE
> THIS IS
> ARIEL ARIEL ARIEL
> RECEIVED MAYDAY.

At this stage the vessel in distress hopefully knows that he is not alone in the world with his problem. You, the acknowledging vessel, will make all speed towards him to render assistance, but you are doing so only because no other acknowledgement was made and you are therefore the only assistance immediately available; however, your yacht at 5 knots may well take some time to reach the distress position. Distress traffic between you and the vessel in distress should always be preceded by the distress signal MAYDAY. You should now endeavour to contact a shore radio station in order to relay the received distress signal: the procedure to relay distress is again in two parts, as follows:

(1) The MAYDAY RELAY CALL:
 MAYDAY RELAY three times
 THIS IS NAME of vessel RELAYING
 three times
(2) A repeated transmission of the DIS-
TRESS MESSAGE:

EXAMPLE

 MAYDAY RELAY MAYDAY RELAY
 MAYDAY RELAY
 THIS IS ARIEL ARIEL ARIEL
 MAYDAY
 FOXGLOVE
 56 DEGREES 20 MINUTES NORTH
 05 DEGREES 10 MINUTES WEST
 HOLED AND ABANDONING TO
 LIFERAFT
 FOUR ADULTS
 OVER.

In UK waters it would be almost, if not
100 per cent certain that a shore station
would acknowledge a distress signal.
Shore radio stations keep a continuous
watch on Channel 16, and they have powerful transmitters and receivers with aeri-
als sited at high elevation. However, it
just could be that your yacht is in a position between the distressed vessel and the
shore station and that you are at the
extreme of his transmitter range, or
indeed that his transmissions are very
weak.

 Assuming that a coast radio station has
acknowledged the distress, the station
now takes control of the situation and
may well impose radio silence on Channel 16 by issuing the following insruction
on an 'all stations' transmission:
 SEELONCE DISTRESS
When complete silence on Channel 16 is
no longer necessary, the controlling station will issue an 'all stations' transmission consisting of:

The distress signal MAYDAY
HELLO ALL STATIONS three times
THIS IS
NAME of STATION in CONTROL
three times
TIME of handing in of message
NAME of vessel in DISTRESS three
times
PRU-DONCE.

EXAMPLE

 MAYDAY
 HELLO ALL STATIONS ALL
 STATIONS ALL STATIONS
 THIS IS
 SOLENT RADIO SOLENT RADIO
 SOLENT RADIO
 1035 FOXGLOVE
 PRU-DONCE.

When all distress traffic is finished, the
controlling station will issue a 'resump-
tion of normal radiotelephony (RT) working'.

EXAMPLE

 MAYDAY
 HELLO ALL STATIONS ALL
 STATIONS ALL STATIONS
 THIS IS
 SOLENT RADIO SOLENT RADIO
 SOLENT RADIO
 1035 FOXGLOVE
 SEELONCE FEENEE.

THE URGENCY SIGNAL PAN PAN

An urgency signal is used whenever a
very important message regarding safety
is to be transmitted, and takes the following form:
 The urgency signal PAN PAN
 transmitted three times

Usually an 'all ships' three times
THIS IS
NAME of VESSEL three times
Nature of URGENCY and HELP required
OVER

EXAMPLE

PAN PAN PAN PAN PAN PAN
ALL SHIPS ALL SHIPS ALL SHIPS
THIS IS PROSPERO PROSPERO
PROSPERO
ENGINE FAILURE ANCHOR NOT
HOLDING BEING DRIVEN ASHORE
TWO MILES DUE SOUTH
HENGISTBURY HEAD REQUIRE
TOW URGENTLY
OVER.

The station issuing an urgency call and message should cancel the same when the action is no longer necessary, by transmitting an all stations cancellation.

The procedures for the important 'mayday' and 'pan pan' calls should be posted next to the radio, and pointed out to the crew. The procedures should be explained by the skipper and if necessary a 'radio off' dummy run made. It must be pointed out that the licensed user, probably the skipper, may be the one for whom help is being sought.

The Emergency Position-Indicating Radio Equipment

This portable electronic aid may well be considered to be one of the most important pieces of safety equipment available to the yachtsman. When activated, it automatically broadcasts its position, and modern EPIRBs also identify the vessel in trouble so that the relevant rescue operation can be mounted. At the present time,

the organization 'Safety of Life at Sea' (SOLAS) is undergoing a massive improvement, and an EPIRB forms but one arm of the 'Search and Rescue' (SAR) part of the new 'Global Maritime Distress and Safety System' (GMDSS). If you should ever have to abandon a yacht, take the EPIRB with you.

The EPIRB should be tested at least once every month (trip), using the 'on board' facility. The expiry date of the battery should be noted, and it should be renewed when necessary.

Life Jackets and Safety Harnesses

These should be demonstrated to, and tried on by all new sailors: even die-hardened members of the crew can benefit from a reminder of how to use such an important piece of safety kit. Once the times of use have been explained, non-swimmers are invited to wear them whenever they feel the need. They are mandatory at night when on deck, in fog and in foul weather.

Safety netting for children. Note the anchor winch – to make a sailor's life easier.

When fitted to crew members in the cockpit, safety harnesses should be clipped to a substantial, purpose-built safety point; when working on deck, the harness should be clipped to the fore and aft jackstays. When moving about on the deck, one's body should be kept low. When working at the mast, the mast itself or the shrouds can be used as safe clipping points.

If both jacket and harness are fitted, the life jacket goes *on top* of the safety harness.

Life-Buoy and Dan-Buoy

These two life-saving pieces of equipment and their joint function are fully described in Chapter 1, but the information is repeated here for completeness of this chapter on safety.

They are normally mounted on purpose-built brackets attached to the push-pit (see Fig. 1b (11)). The dan-buoy is a bottom-weighted telescopic pole with a small flag attached to the top, and weighted so that it floats vertically when heaved overboard. The life-buoy, usually connected to the dan-buoy by a long buoyant line, has a waterproof light attached to it which is held upside down in the bracket and immediately illuminates when placed upright. Whenever the need arises these two pieces are removed from their brackets, the dan-buoy extended, and the buoyant line then uncoiled and both thrown towards the person in the water – and please note, '*towards*' rather than '*at*'.

The buoyant line is an extremely good idea because it lies out on the sea surface, and the person in the water need only swim towards the area between the two pieces of floating equipment so as to make contact with the line before pulling in the life-buoy.

The life-buoy light should be tested at regular intervals because constant exposure to the sea air and occasionally to sea water may cause corrosion.

Grab Bag

As we shall see, a yacht's life-raft contains several items considered essential for people who have to abandon their yacht; however, the very nature of the life-raft means that these prepacked items are kept to a minimum. The yacht's grab bag contains items which the skipper has chosen to complement and extend those provided in the raft. Typically a grab bag will have the following: an EPIRB, extra flares, extra hard-tack and extra water.

Life-Raft

All skippers should attend an RYA 'safety at sea' course, when the content and operation of these devices is explained and practised. The practice usually takes place in the safety of a local swimming pool, with the pupils in full foul weather dress. The experience is well worth having, because swimming in foul weather gear takes some doing, not the least because the sleeves of the jacket fill with water. Note that, unless fitted with a crutch strap, the life jacket or buoyancy aid rides up under your chin.

When the life-raft is launched it is attached to the yacht by a short rope, and the crew should enter the raft by first jumping the short distance into the entrance hole, then climbing in over the edge. Getting out of the water into a life-raft is not easy, however, although under some circumstances it is unavoidable. I was surprised to be told 'let the fittest, strongest person present climb in first', until I realized the invaluable help he affords in helping the not-so-strong to get into the life-raft.

The crew needs to know how to launch the life-raft, how to get into it, and what to do once they are all safely inside. In the unlikely event of abandoning ship, each member of the crew must know what tasks have to be performed:

Who launches the life raft?
Who informs the coastguard?
Who gets the grab bag?
What else needs to be done?

The skipper alone may not be able to do all those things that will need to be done; moreover, he may even be injured or unconscious. It is not the author's intention to frighten the reader, but problems at sea tend to come in threes, each more dangerous than the last. Several books on the market relate the experiences of people who have had to abandon ship, and in several of these cases the time between the initial damage to the vessel and its total disappearance below the sea surface was measured in minutes.

Flares

Flares should be handed round the crew and the user instructions read, and what type they are, how to use them, and their storage place on the boat should be fully explained.

Fog Horn

Where is it fitted, and is it operated by mouth or aerosol? If an aerosol, is it full? Does it work?

Man Overboard

Since each overboard incident is unique, there cannot be one single routine of rescue. However, three very important operations need to be successfully carried out in order to rescue a person who has fallen overboard – although precisely how these operations will be achieved depends on the individual circumstance. First of all, if we visualize a yacht with several experienced crew enjoying a pleasant daylight sail in the Solent on a calm sea, then the following plan may be successful:

(1) 'Man overboard' must be found

If the fall is seen the observer must shout out the fact, and the helm must be put hard over to put the bow through the wind. If the helm is kept hard over the boat will circle indefinitely; the mainsheet should be hauled in to bring the boom in line. A life-buoy or a dan-buoy should then be thrown to fall near the person in the water, and mayday raised on the VHF. With luck the yacht will be circling an uninjured person in the water, who may be brought back to the side of the vessel with the aid of a thrown line.

In some cases the yacht may not circle indefinitely and it may be necessary to sail a short distance away on a beam reach, and to tack and return on a reverse beam reach, stopping the boat alongside the person in the water by heading into wind. Sailing away like this gives the skipper time to settle the boat and crew down whilst detailing the plan of rescue. This 'sail away' routine may be improved by going off on a broad reach, say 120 degrees to the wind, then tacking to return on a fine reach of 60 degrees. This

107

should allow the crew to spill wind from the jib and main in order to slow the boat down on its final approach to the person in the water.

If the person overboard is injured, unconscious or lying face down in the water it may be necessary for another crew member to be lowered over the side in order to bring the injured party back to the boat. It is, I hope, obvious that deliberately lowering another person into the sea is fraught with added danger. The yacht must be stopped in the water and all sail dropped, with no ropes over the side and the engine running but the prop not engaged. The person who is lowered – well away from the propeller – must be fit, adequately dressed and well secured to the yacht.

(2) 'Man overboard' must be got aboard

With the person(s) alongside and plenty of sea room, the engine should be switched off. Where possible, use the stern ladder or a portable boarding ladder over the side. The spinnaker halliard, a handy-billy and chest strop could be used to get people back on board (see Fig. 83).

If the fall itself was not noticed – ie at some time later a crew member is found to be absent – then a search pattern needs to be carried out and the appropriate radio message transmitted to the coastguard.

(3) Any trauma must be relieved

Once securely inboard, the rescued person must be dried off and rested. Even if

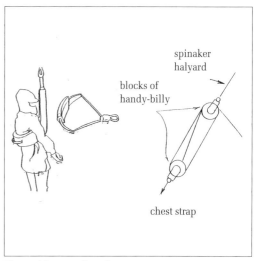

Fig 83 Man overboard recovery.

the coastguard was not previously aware of your drama, he must now be informed and medical advice sought.

We began our discussion on man overboard with certain assumptions: nice daylight sail, experienced crew and a calm sea. 'Experienced crew' implies crew members who are well practised in the above routines, and a skipper who knows exactly how his boat responds to sail trim. You may well ask how a member of such a team and in such lovely sailing conditions could fall overboard – and the answer is that such things do happen. However, it is much more likely to happen to inexperienced persons and generally in much worse conditions than those assumed above.

'Man overboard' is a highly dangerous situation, and rescue is by no means certain. BE CAREFUL: don't go overboard in the first place.

12
WEATHER

In previous chapters we have dealt with some of the many skills encompassed by the term 'seamanship', that technical skill and ability which must be acquired before taking a yacht and its crew away from a safe berth. In addition to all of that, a skipper must have a good understanding of how the local weather may develop and change in the immediate future. In harbour the decision of when to start the trip, if at all, will depend to a very large degree on his/her forecast of likely wind and weather states. Observations whilst at sea will either allow the cruise to continue, perhaps under shortened sail, or determine that a run to the nearest safe harbour should be made.

Such skills as these decisions depend upon are not easily acquired, weather forecasting is not an exact science, and sooner or later unexpected bad weather will be experienced by all who go down to the sea ... It follows that we should always be prepared for the onset of bad weather, and, given some knowledge of weather prediction and visual evidence that a situation may worsen, we can perhaps ask and answer questions such as: Are the sails suitably set? It is easier and safer to leave harbour under-canvassed than it is to change and/or reef sails in

heavy weather? Is everything stowed away down below? Are the crew suitably attired? Have they eaten? Is emergency food prepared? Are the crew sufficiently experienced for the expected weather? If all the answers are affirmative, then a tentative trip to the harbour/river mouth might be made: a fast sail to windward in a good blow takes some beating!

Reading this chapter on weather will not make you into a meteorologist; the subject is far too complex, and weather patterns far too variable for a book of this type to make any such claim. However, the prudent skipper, having spent some time studying how weather patterns form, will also have been listening to and watching weather forecasts, over several days, prior to a planned sailing trip. He will know full well that sitting at a safe upriver berth can give little indication of the weather in open water.

Weather: the Basics

The energy driving the earth's great weather systems is derived from the heat of the sun. At the equator, the direct rays of the sun warm the earth and indirectly the sea above it; the diurnal spin of the globe allows the equatorial lands to cool at night, only to be heated again the next day, whilst the temperature of the equatorial seas and oceans tends to remain constant over this same twenty-four-hour period.

At higher latitudes the sun strikes the earth less directly, its heat energy being spread over an increasingly large area until at the poles more heat is lost by radiation from the surface than is gained from the sun. It is this differential heating of the earth's surface which generates high and low pressure air systems and both warm and cold sea currents.

To begin with we will examine the composition of the air in which we live and of which Aristotle said 'It is absurd to suppose that the air which surrounds us becomes wind simply by being in motion.'

The Atmosphere

The atmosphere covers the entire world, and we live at the bottom of this sea of air, at an average pressure of some 1,013.5 mbs. Air pressure is quite literally the weight of the air particles above the earth's surface; it therefore falls with increasing height, but, even at 100 miles (160km) above the earth, there is still sufficient air density to bring meteors to a white heat.

The lower part of the atmosphere, called the troposphere (see Fig. 84), varies in height from 5 miles (8km) at the poles to 10 miles (16km) at the equator. It contains the greater mass of air and most of the atmosphere's water vapour, and, very importantly, the temperature of the air within it decreases uniformly with height. All the world's weather systems develop in this lower region of the atmosphere.

At the upper level of the troposphere the fall in temperature with increasing height ceases, and the transition can be very abrupt; the region in which it happens is called the tropopause. Above the tropopause is the stratosphere in which the amateur meteorologist has little interest.

The composition of the atmosphere is quite complex, but fortunately we are only interested in its temperature, its density or air pressure, and the amount of water vapour present. In fact the amount of water vapour in the air is a function of the air temperature, in that the higher the air temperature the higher the amount of water vapour that can be held. When a body of air is cooled, the water vapour condenses into water droplets, and if cooled sufficiently precipitation occurs.

In equatorial regions the surface air in contact with the sea, oceans and land is heated by conduction, and this contact heating causes the surface air to absorb, by evaporation, differing amounts of water vapour depending on the type of

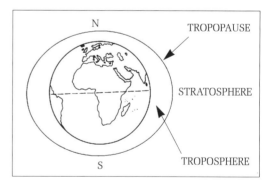

Fig 84 The earth surrounded by the troposphere. The troposphere contains almost all of the water vapour and air molecules, it is some 10 miles deep at the Equator, falling to 5 miles deep at the poles. Temperature in the troposphere falls with increasing height.

surface below it; up to 4 per cent of an air/water mixture may be water in suspension. It follows that the heated surface air will expand, become less dense than the air above and therefore rise through it. This phenomenon is worthy of further deliberation, in that the surface air may be lying over deep or shallow sea, or it may be over land of differing terrain – sand, rock, black earth or green foliage. These differing terrains may extend over several square metres or over several thousand square miles, and the air in contact with a particular terrain will have a particular temperature, pressure and water vapour content.

The result is that individual localized bubbles of heated air will rise, independently of other localized bubbles of air. The rising bubbles may be a few miles or a few thousand miles in diameter, and are generally refered to as 'air masses'. Whether bubbles or air masses, it is important to grasp that they will be at different temperatures, they will differ in air densities, and they will hold different amounts of water vapour. The rising air masses will be cooled by the upper layers

of colder, more dense air through which they are rising, and several results are possible:

(1) The rising, cooling air mass will reach a height at which its pressure becomes equal to that of the air surrounding it. At this point it will cease to rise, it will be pushed to one side by warmed air following up behind it, and will fall back to the surface to be reheated, thus restarting the cycle all over again.

(2) The rising mass will reach its dew point, and the water will condense out as water droplets, becoming cloud or rain.

(3) The rising, cooling air will reach the tropopause without its pressure falling to that of the surrounding air. In this case the air mass will split, some of it moving in a northerly and some in a southerly direction at high altitude, again being pushed aside by following rising air masses.

Fig. 85 illustrates these three effects. It is the last possibility that is of interest, not least because we have left a partial vacuum at the sea surface; this will of course

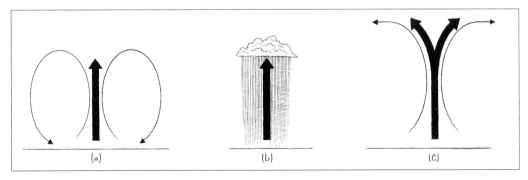

Fig 85 (a) The warm air mass rising and cooling until its pressure equals that of the surrounding air. It is pushed aside by following rising air, where it falls back to earth to begin the cycle all over again.

(b) The warm air mass cooling and, as it rises reaches it's dew point, precipitation occurs.

(c) The warm air mass, reaching the troposphere still at lower pressure than surrounding air, moves poleward.

be quickly eliminated by local surface air moving in to fill it. The new surface air will become heated, will absorb water vapour, fall in pressure and rise to the tropopause in its turn.

In both hemispheres the high altitude air masses, each with their own characteristics of pressure, temperature and water content, will migrate towards the poles. This migration away from the equator and the consequent cooling of the air mass will cause the high-level air masses to become denser than the air below as they move polewards. The result of this cooling is a gradual fall from high level to surface level, and once at the surface the air will move back towards the equator: this completes a circulatory air cell in which the heated rising air at the equator moves polewards, becomes cooled, falls back to the surface and returns to the Equator.

Air Movements on a Rotating Earth: Coriolis

Unfortunately the simple one-cell view of upper and lower surface air masses moving north and south between the poles and the equator does not agree with observable air-mass movement. The rotating earth causes air streams to turn to the right in the northern hemisphere, and to the left in the southern hemisphere (see Fig. 86); this is known as the Coriolis effect, and can be explained as follows:

Consider that the heated air mass at the equator forms a single ring of warmed air around the entire earth. The earth's speed at the equator is approximately 1,000mph (1,609km/h) and our ring of surface air is moving at this same speed. The ring of air rises, reaches the troposphere and moves

polewards, and as it does so, its speed becomes increasingly fast relative to the air and earth surface below it; this is because the earth's surface speed, west to east, reduces with increased latitude.

The upper air still travelling at 1,000mph would therefore have a west-to-east velocity component added, such that its initial poleward direction is deflected, to the right in the northern hemisphere and to the left in the southern hemisphere. The effect increases with increased latitude so that ultimately, at approximately 60° of latitude, the upper air stream becomes the high altitude jet stream travelling west to east at high speed. The upper air stream also cools as it moves away from the equator, becomes denser than the air below it, and at approximately 30° latitude some of the upper air stream falls back to the earth's surface.

At the surface this air mass splits, some of it moving back towards the equator and some moving polewards. The former passing over an ever-increasing earth radius, acquires an east–west velocity component and in the northern

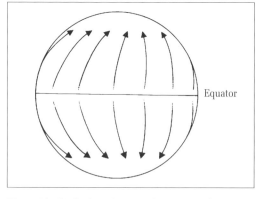

Fig 86 The high altitude air at the Equator, having reached the troposphere moves north and south, the curvature of its path being due to the Coriolis effect.

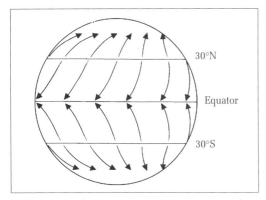

Fig 87 The surface air at 30°, some returning to the equator to complete the basic circulating air cell, the remainder moving polewards. Both are subjected to Coriolis. In the northern hemisphere the returning southerly wind becoming the NE trades. The pole-going air mass becoming the SW winds so often experienced in the UK.

hemisphere becomes the north–east trade winds. That air moving polewards, on the other hand, passing over an ever-decreasing radius, acquires a west–east velocity component and in the northern hemisphere becomes the south-westerlies. The surface winds caused by this effect are shown at Fig. 87.

Atmospheric Pressure

The earth's mean standard atmospheric pressure is approximately 1,013.5mbs. However, it does vary with latitude and in different areas of the same latitude. The Equator, for example, with warmed surface air becoming less dense and rising, is an area of relatively low pressure. The differing surfaces in this region – deep and shallow areas of oceans, the different land surfaces, each at its own temperature – will give rise to equatorial air masses all at differing but generally low pressures.

We have seen earlier that at approximately 30°N latitude the cooling upper air masses, having become denser than the air below, fall back to the earth's surface. In doing so, the increasing density (pressure) on these same masses causes their temperature to rise such that not only will they hold on to their own water content, but they will also absorb any surrounding water vapour – that is, they fall to the earth's surface as dense, warm, dry, desiccating air masses. At the surface some of the air flows towards the equator becoming the north–east trade winds, and some moves polewards becoming the south-westerlies.

We have seen how the equatorial region as a whole may be considered an area of low pressure, generating individual air masses which have their own unique temperature, pressure and water vapour content. The polar regions, on the other hand, are areas of high pressure: only a small amount of heat from the sun reaches the polar surfaces, and consequently more heat is lost by radiation at all height levels than is gained from the sun. The result is that high level, very cold, very dense, very dry air continually descends in these regions, and at the surface, high-pressure air masses migrate towards the equatorial regions.

In moving over a rotating earth's surface of increasing radial distance, these cold air masses will be given an east-to-west velocity, thus becoming in the northern hemisphere the polar north-easterlies. At approximately 60°N the north-easterly polar air masses will meet the relatively low pressure south-westerly air masses. The result is that the warm, comparatively low pressure southerly air masses will tend to rise up over the cold, high-pressure north-easterly air masses, creating an area

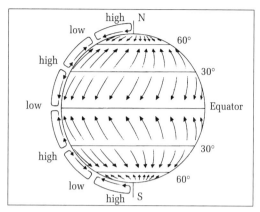

Fig 88 The very simplistic view of global air masses. In the model, low pressure zones exist at the Equator and at 60° latitude. In both these areas warm air is rising from the surface. At 30° latitude and at the poles air is constantly falling to the surface creating high-pressure air masses.

of low pressure at approximately 60°N. Fig. 88 shows the planetary wind systems generated by the movement of air masses over the earth's surface.

Atmospheric pressures taken simultaneously over a given area and marked on a chart allow those points of equal air pressure to be joined; the lines of equal pressure are called isobars, and resemble contour lines used to depict hills and valleys on land maps. Fig. 89 shows the isobars surrounding a high-pressure air mass; isobars are normally shown at 2 or 4mb intervals. The centre of a high-pressure air mass will be ringed by isobars which decrease in pressure with increasing distance from the centre; the centre of a low-pressure air mass, on the other hand, will be ringed with isobars showing increasing pressure with increasing distance from the centre.

Atmospheric pressure around the world varies between a lower level of 950mbs and an upper level of 1,050mbs, with a standard pressure of 1,013.5mbs.

Unfortunately once again our picture of a ring of warmed air rising at the Equator and moving uniformly north and south as high altitude air, eventually to form the planetary wind systems of Fig. 88, is a little too simplistic to explain the many variable weather systems to which, for example, UK waters are exposed.

We need to note that the northern hemisphere is not a homogeneous whole, but has large land masses separating several oceans and seas. Some of these land masses are permanently covered in ice and snow, others are not, but may experience large changes in temperature over the winter and summer periods. Subtropical land masses remain relatively warm throughout the year. Moreover land-mass height above sea level varies enormously, some masses having areas of great height; equally the seas have areas of greatly varying depths, and they contain both warm- and cold-water currents such as the warm Gulf Stream and the cold Labrador Current.

A consequence of these and other features is that permanent and semipermanent areas of high and low pressures exist in both hemispheres, and it is in these regions that air masses are born, each with their unique characteristics of temperature, pressure and humidity. In the northern hemisphere high pressure zones are found over the Azores, Continental Africa, Siberia and the Polar regions. The Azores high is permanent, but most of the others vary in intensity between summer and winter, and for example the Siberian high of winter becomes the continental low of summer. Other sources of low-pressure air masses are inland waters in winter, such as the Mediterranean, Black and Caspian seas. Fig. 89 shows the region of the Azores high-pressure area.

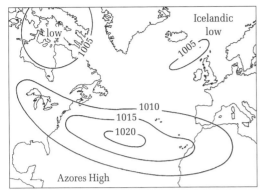

Fig 89 The permanent Azores high pressure region. Most of the warm wet air masses moving across the UK are generated on this area becoming the UK's south-westerly air mass.

The Azores, Polar and Siberian Highs

The Azores high is a permanent area of high pressure. Air masses moving up from the Azores over relatively warming seas become less dense and consequently absorb sea water as they move northwards. These air masses arriving over the UK are identified as WARM, HIGH PRESSURE and WET. Please remember that individual masses have their own characteristic values, and tend to remain as independent entities having their unique temperature, humidity and pressure.

The Polar and Siberian highs are also areas of high pressure, being sources of air masses whose characteristics are very cold, very dense and dry. In moving south over warmer terrain they may become less cold and therefore less dense; however, when compared to an air mass originating from the Azores or the African

continent they are identified as COLD, HIGH PRESSURE and DRY.

Air masses are classified as follows:
(1) maritime polar (mP)
(2) continental polar (cP)
(3) maritime tropical (mP)
(4) continental tropical (cT)

Fig. 90 shows the path taken by these air masses when moving towards the UK.

Air Movement within Air Masses

We have already seen that air within a low-pressure air mass is continually rising from the surface and must be supported by an inward flow of air or horizontal convergence in the lower atmosphere. The air in a high-pressure air mass is continually falling to the surface and must be associated with outflow or horizontal divergence of air in the lower atmosphere. The vertical movements of air are very gentle, being only a metre or so in a day. Fig. 90 illustrates the vertical and horizontal air movements within air masses.

The speed of the horizontal flow of air into or out of an air mass – convergence or divergence – is a function of the pressure gradient force within the air mass: the higher the pressure gradient, the higher will be the speed of the air flow. Convergence or divergence, both air flows are also subjected to the Coriolis force, and in the northern hemisphere the horizontal flow will turn to the right. Ultimately the two forces will balance and the air flow will move parallel to the isobars. Fig. 91 illustrates the result of the Coriolis effect upon high-level air.

A further complication occurs below heights of 5/8m (1km) where the air

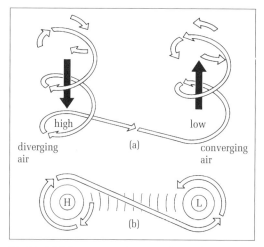

Fig 90 A high-pressure air mass, air moving in at high altitudes, is affected by Coriolis, eventually moving clockwise parallel to the isobars. At the surface the air is forced away from the centre by more air following it down, it therefore cuts across the isobars as it moves away under a pressure gradient towards a centre of low pressure. In moving over the surface the resulting friction slows the air movement, Coriolis is reduced, the pressure gradient becomes the greater force, causing the air to move inwards towards the centre of the low.

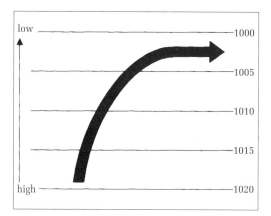

Fig 91 High level air above 500 metres unaffected by surface friction moves under a pressure gradient from a high to low pressure area; however this moving air is subjected to Coriolis, turning to the right in the northern hemisphere. The two forces, pressure and Coriolis balance at the point where the air is moving parallel to the isobars.

particles are affected by surface friction; the effect is to decrease the wind speed so that a new equilibrium is established, with the lower air moving across the isobars towards the lower pressure. Over the sea the lower air cuts the isobars at approximately 15°; over land where the friction is higher the angle is nearer 30°. Fig. 92 illustrates how surface friction over the sea affects the direction of low-level air flow between a high- and a low-pressure air mass.

Figs. 93 and 94 show a high- and low-pressure system moving across the UK, and the direction of the resultant surface air in the channel.

A basic picture of the equatorial- and polar-surface air masses which affect the UK can now be summarized:

(1) A polar-sourced surface air mass is cold, dense, of high pressure and dry. It moves in a south-westerly direction and is seen in the UK as north-easterly winds.

(2) A tropical-sourced surface air mass moving up from the Azores, latitude 30°, is warm, of high pressure* and wet. It moves in a north-easterly direction.

The Polar Front

Fig. 95 shows a cP air mass, moving south-westerly, meeting an mT air mass moving north-easterly. If these air masses are small they may pass one another with little effect; however, if they are large and of very different characteristics, their meeting can and very often will produce a frontal depression or even a family of such depressions. The line along which these two differing air masses meet is called the polar front.

*Although tropical-sourced air masses from the Azores are high-pressure ones, they are much less dense than polar sourced air masses.

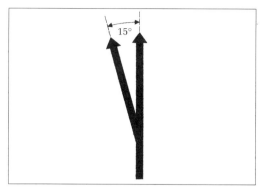

Fig 92 Surface air molecules are subjected to friction; the air mass moving down the pressure gradient slows. Coriolis effect is reduced, the air backs by some 15°.

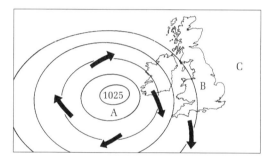

Fig 93 The high-pressure air mass moving slowly eastwards across the UK. When the centre of high pressure is at position A, B and C the winds in the Channel will be northerly, easterly and southerly respectively.

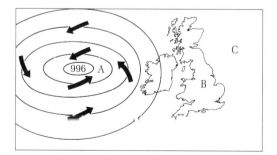

Fig 94 The low-pressure air mass moving across the UK, at positions A, B and C will produce winds in the Channel of southerly, westerly, and northerly respectively.

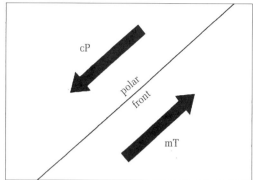

Fig 95 The meeting of two opposing air masses; the features of cP being dense, high pressure, dry and cold will not easily mix with mT whose features are lower pressure, wet and warm. The junction is called the polar front.

Frontal Depressions

A surface of separation called the polar front exists at the boundary where the two air masses meet; very little mixing will take place at this junction because the temperature, pressure and vapour content of the two masses are so different. However, at the surface of separation the cold, dense cP will undercut the warm, less dense mT, and the mT will ride over the top of the cP.

The exact cause of this twist, or perturbation, at the polar front is difficult to pin down; it may be simply a matter of two masses of air particles which in coming together will not mix, but in their passing set up such a friction that eddies are formed forcing the undercutting by the cold and the overriding by the warm. Figs.96a and b show the vertical and horizontal views of the initial wave.

If there is sufficient energy stored in the two air masses, the wave will develop along the lines shown at Figs. 96 c, d and e. At c, the warm air is twisting anticlockwise, sliding over the cold, and, by raising the temperature of the cold air at the surface, it is also producing a slow-moving, surface WARM FRONT pushing into the cP. At the rear of the wave the cold air continues to undercut the warm just by forcing the warm air vertically upwards; the result of this action is to extend a fast-moving cold front into the mP air mass. The developing depression moves parallel to the isobars, generally in a north-easterly direction, and continues to rotate anticlockwise.

At Fig. 96d the faster-moving cold front has caught up with the warm at the centre of the depression, so warm air is being gradually lifted from the surface, and the combined single front is called an occluded one. Fig. 96e shows a completely occluded front, where all the warm air has been lifted from the surface, and the depression will fill and dissipate.

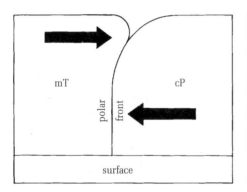

Fig 96(a) The mT air mass is shown moving over the cP. The cP is shown under cutting the mT.

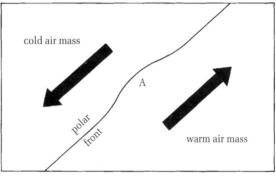

Fig 96(b) Looking down on to the front, the small initial perturbation is shown as mT rides over the top of cP. The overall air pressure at this point 'A' is reduced.

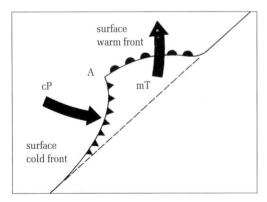

Fig 96(c) The developing depression centred on the low pressure point 'A' has several points of interest. The mT air mass slides up and over the cold cP and at the same time produces a warm front at the surface by literally heating the cold cP air mass. The warm front continues to develop as a warm arc pivoted at point 'A'. At the cold front the cP air mass undercuts the warmer mT literally forcing this lower pressure air up and out of the way. The surface cold front is also a moving arc pivoted at 'A'. The cold front moves much more quickly than does the warm front; this is because the warm front must heat the adjacent cP as it pivots, a slow process. Whereas the cold front advances by simply lifting the warmer air ahead of it, up and out of the way. Notice that the symbols identifying the warm front are nice friendly curves, while the cold identifiers are nasty sharp spikes.

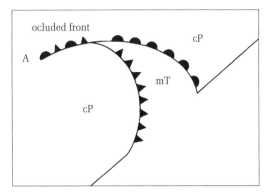

Fig 96(d) The faster moving cold front has caught up with the warm one at the centre of the low pressure area, the low pressure is lifted off the surface at 'A'. Notice that the whole depression has rotated anti-clockwise about point 'A'.

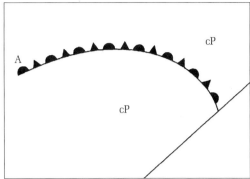

Fig 96(e) The depression is completely filled, all warm air is lifted from the surface, the area of low pressure centred on 'A' has been completely filled.

The winds associated with a frontal depression are shown at Fig. 97. The frontal depression is shown moving north-easterly with an observer at 'A'. The arrows indicate the surface wind, turned in towards the centre of the low at some 15° to the isobars. The spacing of the isobars indicates the wind strength, and as we shall see later, can be translated into the wind speed. The whole developing cyclonic system is moving in a generally NE direction with the two fronts rotating anticlockwise about the central low-pressure area.

At some distance from the approaching front, position 'A' will experience a southerly wind. Immediately before the warm front, the wind may back as the depression continues to rotate anticlockwise around its low centre; wind backing immediately ahead of a warm front is very slight, and is not usually noticed by an observer. However, the passing of the warm front is often accompanied by a large wind veer, usually very noticeable. The passing of the cold front is also associated by a large and noticeable wind veer.

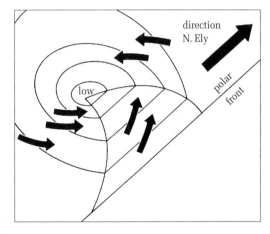

Fig 97 The wind within the depression cuts the isobars as shown; please notice how sharply the wind backs at both the warm and the cold fronts.

Cloud Formation

In discussing the cloud formation associated with a frontal depression, we need to recall that the air in the warm sector originated in the Azores high, and that it began moving in a north-easterly direction as a warmish, rather high-pressure, dry air mass. Moreover in moving over sub-tropical water its temperature and therefore vapour content may have increased and its pressure reduced.

In Fig. 96c, air from the warm sector is seen sliding up over the cold, dense air of the cold sector, while at the same time causing the surface warm front to move into the cold sector. The sliding warm air will cool as it rises over the cold air mass, losing its ability to retain water in evaporated form; water droplets will condense, clouds will form and eventually precipitation will occur. The warm air will continue to slide up over the cold, and vapour will continue to condense at increasing height levels, forming the cloud masses shown at Fig. 98. Notice that very high cirrus cloud may be observable some 200 miles (320km) ahead of the warm front.

Fig. 99 shows how the cold-sector air undercuts and literally lifts the warm air very abruptly. Precipitation some fifty miles (80km) ahead of the cold front can be extremely heavy, though fortunately it can often be seen approaching, one clue being the large, dark cumulus cloud formation with the 'anvil'-shaped head.

Cloud formation in the warm sector is determined by the characteristics of the air mass. If this mass has been cooled by passing over north Atlantic and western approach waters, it may be approaching its dewpoint, at which time visibility will be hazy, clouds may form and drizzle or rain may occur.

Changes in air temperature are associated with frontal depressions. The observer at 'A' in the cold sector will notice an increase in temperature as the warm front passes, when the air may feel muggy and warm, whereas the cold front will bring a drop in temperature and the sky should become bright and clear as the cold, dry air passes over. Changes in air pressure are often associated with the passing of a frontal depression. The air pressure should

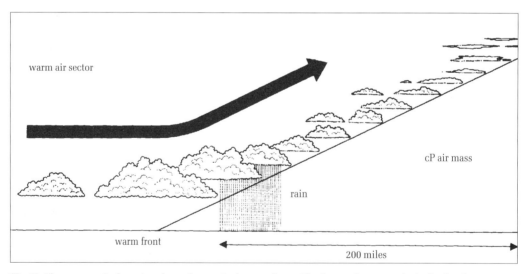

Fig 98 Shows a vertical section through a typical warm front. The front advances relatively slowly. Air in the warm sector rises up over the cold cP air mass, in doing so clouds usually form as the temperature of the rising air falls. Precipitation is generally associated with the area ahead of the warm front; how much cloud? how much rain? depends on the individual air masses.

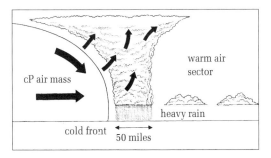

Fig 99 Shows a vertical section through a cold front; the rapid rise of the warm sector air ahead of the front generates storm clouds causing very heavy rain over a short distance, 50 miles, ahead of the front.

fall steadily as the warm front approaches; it will remain steady in the warm sector, then rise abruptly as the cold front passes, and again remain steady.

Land and Sea Breezes

During a warm, sunny day the land heats up much more rapidly than the sea, and at night it cools much more quickly; it can take up to five times more heat energy to raise a unit volume of sea water through the same temperature change as a unit volume of sand. On a warm, sunny day the difference in land/sea temperatures becomes significant in the late morning as the sun heats the land; the temperature difference will fall to zero in the late evening; then during the night, as the land cools rapidly, its temperature may fall below that of the adjacent sea.

The heated land warms the low layers of air above it, the warmed air rises, and the cooler air moving in from the sea as a converging balance to replace it is recognized as a 'sea breeze'. The warm, upper air over the land will move out over the sea, and as it cools it will fall thus completing a closed air circulation as shown at Fig. 100. If the air over the land is sufficiently wet, cumulus clouds will form over the land.

A sea breeze will increase in strength during the day, and its effects may overpower the main weather system in its

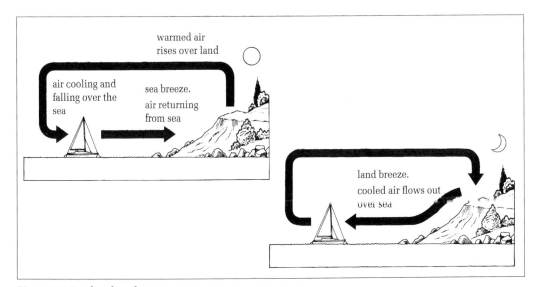

Figure 100 Land and sea breezes.

immediate area so that it becomes the significant wind force for several tens of miles out to sea. Coriolis effects will turn the sea breeze to the right, and by late afternoon the breeze could be running along the coast. Towards sunset, as the power of the sun fades away, the sea breeze will reduce and stop.

In the evening and during the night the temperature of the cooling land will fall below that of the sea, and this will reverse the air-pressure gradient, setting up a land breeze over the sea. The night-time temperature difference between the land and sea is usually much less than during the day, resulting in a comparatively weak land breeze.

Cloud Types and Formation

Cloud types are defined by their shape and altitude. Three general types are identified: the cumuliform or heaped, the stratiform or layered, and the high-altitude cirrus or ice cloud.

Low-Height Clouds

Clouds forming below about 1½ miles (2km):
(1) *Cumulus (Cu)*: small, cotton wool-like clouds with a flat base, sometimes called fair weather clouds, often forming along the coast as the land is heated. When seen at sea, they may well indicate the presence of land over the horizon.
(2) *Stratus (St)*: the whole sky may be covered in this low-lying, sheet-like layer of cloud. In the presence of an mT air mass it may herald the formation of sea fog.
(3) *Stratocumulus (Sc)*: this is a combination of the above two types; it may

transform into stratus, and may well cover the whole sky.

Medium-Height Clouds

Clouds forming between 1½ and 4 miles (2 and 6 km):
(1) *Altocumulus (At)*: small groups or lines of cloudlets of cumulus, covering large areas of the sky; individual groups may show shadow, and at some angles of the sun be tinted red, green or yellow. The term 'mackerel sky' is given to very small cloudlets.
(2) *Altostratus (As)*: a stringy grey or bluish-grey cloud, having the appearance of a watery sky; the outline of the sun appears blurred through this cloud. It often precedes a warm front and warns of rain from following nimbostratus.
(3) *Nimbostratus (Ns)*: this is a stratus form of rain cloud, usually dark grey in colour and often seen with curtains of rain below it. It may be very thick. It is the middle-layer cloud of a frontal depression, and precipitation is often continuous.

High-Height Clouds

Between 4 and 9½ (6 and 15km):
(1) *Cirrus (Ci)*: this is an ice cloud often of 'mares'-tail' appearance. Dense banners of this high cloud may indicate very bad weather to follow.

Two major rainclouds, nimbostratus and cumulonimbus, may span across all levels. Also, the big thunderhead cumulonimbus associated with cold fronts often stretch to the tropopause where the cloud spreads to form the 'anvil'.

There are many, many variations of the above descriptions. It should be remembered that any air mass movement which leads to the ascent of the air within it to the level of its dewpoint will cause cloud to form and possibly precipitation.

Fog

The sailor's fog is generated by one of two mechanisms: advection fog which originates at sea, and radiation fog which, although originating on land, may well roll out over the adjacent sea.

Sea Fog

A warm, wet mT air mass moving into the north Atlantic towards the British coast may well pass over a much colder sea surface, particularly in spring and early summer. In passing over such a cold sea the lower air mass could cool to its dewpoint, and water droplets condense onto salt crystals in the air causing fog, or mist. Fog is present if visibility falls to less than 5/8 mile (1km), whereas mist reduces visibility to between 5/8–1½ miles (1–2km). Sea fog may last for several days: the cooled surface air mass is very stable, it is colder and denser than the air above and so cannot rise through it. Sea fog may persist until the air mass causing it is replaced by an air mass with different characteristics.

These definitions are somewhat academic to the yachtsman, since poor vis. is poor vis. For yachts crossing the channel, for instance, any reduction in visibility is a time for extreme caution, the large amount of merchant shipping in this area, particularly the vessels in the shipping lanes with their rumoured penchant for unreliable watchkeeping, will keep any skipper on his toes. Deteriorating visibility in fog-forming conditions means everybody up and dressed, good watchkeeping, quietness on the boat, a review of 'What if', and, if about to cross a shipping lane, a probable VHF consultation with the coastguard.

Land Fog

Fig. 101 shows high-lying land adjacent to the sea. During a cold night, particularly one following a warm day, the land, and especially high-altitude land, will cool very rapidly, the air above cooling rapidly with it. Water droplets will condense onto sand or dirt particles in the air, forming land fog. The cooled air will become much denser than surrounding, lower-lying air, and will literally fall down the hillside and roll out across the sea. Land fog may even form over low-lying ground, the essential ingredients being a rapid cooling of that land so that the air column immediately above the land is cooled, then fog is formed, spreading out across the surrounding area and the adjacent sea.

Fortunately the following day's sun will usually remove the cause of radiation fog by reheating the land. The unwary sailor moored alongside a harbour wall can be easily caught out by the early morning dissipation of land fog. A clearing sky over the land may occur much earlier than out at sea where the fog may persist for some hours.

Weather Broadcasts

There are many sources of weather information, but the main ones are these: (1) Radio: national and local radio stations transmit weather forecasts at regular

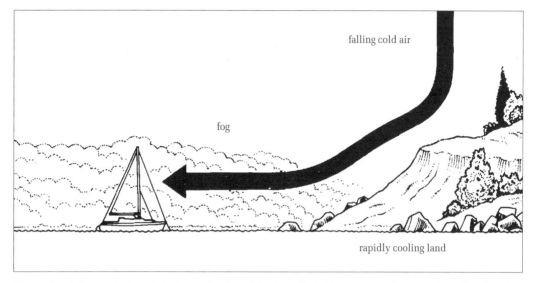

Fig 101 Land fog. Rapidly cooled air over land at night, literally falling out over the sea and condensing, the water molecules condensing onto salt or dirt particles to form land fog.

intervals. The national weather forecasts on Radio 4 are very detailed, indicating the type of weather to be expected in all UK waters over the subsequent twenty-four hours. Local radio stations, on the other hand, give details of expected weather for the inshore waters of their local area. The coastguard stations also issue regular weather information and give strong wind warnings when appropriate.

(2) National newspapers and television companies provide daily weather forecasts, and, though you may not have TV on your yacht or an obliging offshore newsagent, they are extremely good facilities to make use of in the few days before you go sailing. Obtaining a sequence of developing weather patterns before you sail is of enormous help when you are compiling your own weather plot from radio transmissions.

(3) Marinecall: a very useful and immediately available source of weather forecasts. Marinecall divides the UK into a number of areas, identified by telephone or fax numbers. With the appropriate inboard equipment this facility is available twenty-four hours a day.

Onboard Weather Forecasting

The long-distance offshore sailor should practise the skill of weather forecasting. This skill may be the ability to produce his own weather map from a radio forecast and its subsequent interrogation for likely weather states, or it may be the interpretation of a faxed weather map. In either case the prudent skipper will realize that the broadcast itself is only a forecast and not a guarantee of future weather, and that his own predictions are at best guesstimations.

13

IRPCS

The 'International Rules for the Prevention of Collisions at Sea' constitute both a responsibility and an obligation on all mariners who travel upon the high seas and navigable waters connected to those seas. Whenever two vessels are heading for the same piece of water, the IRPCS identifies which of these two vessels shall carry on towards that converging point, and which shall give way. A vessel's priority in such situations is in general defined by her type, her size, or the task she is performing. For example, power vessels give way to sailing vessels in most but not all situations. Very large vessels with deep draught would have a high level of priority in comparison to a small sailing vessel when both are manoeuvring in a narrow channel; on the other hand, in open waters with plenty of sea-room the power-driven vessel is obliged to give way to the small sailing vessel.

For the observance of these rules, all vessels and therefore their priority of passage are identified by one of three systems:

(1) In clear daylight: identification is by means of special shapes carried in the fore-part of the vessel in such a position that they can be seen clearly.

(2) In fog: when visibility is reduced, identification is by means of sound signals such as fog-horns, whistles or bells.

(3) During night hours, from official sunset to official sunrise, or during the daytime in dirty weather: identification is by means of lights, the colour and disposition of which identifies the type, the size and the task of the vessel.

The rules are necessarily comprehensive, attempting as they do to cover every possible sea-going situation and circumstance. Fortunately, unlike the professional navigator, the yachtsman is never required to quote the rules verbatim; however, he does need to know them well enough so that in any given situation at sea he can identify without fail the priority of passage of all other vessels in his area, and can recognize when a collision situation is likely to develop.

Based on this important knowledge of the IRPCS, if a vessel has the higher priority she is obliged to 'stand on' – that is, to maintain her present course and speed – until all possibility of a collision has passed. If, on the other hand, a vessel has the lower priority, then she must 'give way' – that is, change course or speed or both – until all possibility of collision has passed. It is obvious from the above that a good 'look out' all round the horizon is at all times highly desirable, if not essential. The risk of collision must be established as soon as possible, priorities established by both vessels, and the necessary action implemented.

When avoiding action is taken by the give-way vessel it shall be performed as soon as the risk of collision is recognized, and such action should be large enough for the stand-on vessel to see that you are complying with the rules – that is, a big alteration to the course or speed should be made, and these changes should be made in a good, seamanlike (sensible) way.

Fig. 102 shows two power-driven vessels – power-driven according to the rules meaning a vessel under engine. The vessels are heading for the same point, and the rules, as we shall see later, state that, in the situation depicted, vessel A is the give-way boat and vessel B is the stand-on vessel. Vessel A may carry out her obligation in one of two ways: first, she could simply stop in the water or alter course to port and reduce her speed, allowing vessel B to pass before resetting back to her original course and speed. Or – and this is her better option – she could so alter her course that she points to a

position slightly astern of B, maintaining that clearance round B's stern until all danger has passed. Such a move by A would be obvious to B's skipper and it shows a clear intent.

The question must now be asked, what happens in a collision-converging situation when the give-way vessel does not appear to be carrying out any avoiding action? The rules allow for this by the following change of emphasis: the skipper of a stand-on vessel, seeing that the give-way vessel has not taken any avoiding action, 'may take action to avoid collision' (this quotes the rules). If, however, at this stage the stand-on vessel does not take action and the situation worsens to the point where whatever the give-way vessel alone now does a collision is unavoidable, the stand-on vessel 'shall take such action as will best aid the avoidance of a collision'.

The change from 'may take action' to 'shall take action' puts a severe responsibility on the stand-on skipper, since any change to course or speed of the stand-on vessel must be done in such a way that if, at this very late stage, the give-way vessel takes avoiding action, the stand-on's action does not worsen the situation.

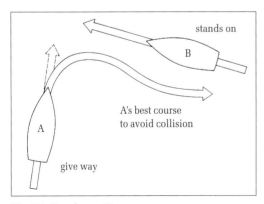

Fig 102 Stand on – Give way.

Anticipating Collision Situations

The avoidance of collisions with vessels ahead or astern is best achieved by good visual observation. The avoidance of collisions with vessels which are off to one side or the other and which may be on a converging course is best achieved by using a hand-bearing compass: if the bearing of the other vessel is constant or nearly so, then a collision will occur unless avoiding action is taken. It may be that the first bearings taken of the other vessel indicate that all is well; however, since there was sufficient concern to take the bearings in the first place, it would be prudent to repeat them until all possibility of collision has passed.

There are two other ways that the possibility of collision may be checked:

(1) Line up the other vessel with two fixed objects on your own boat – for instance, if a coach roof winch, a shroud and the other vessel are in line and remain so, a collision will occur. If these three items do not remain in line, all is well, and course and speed can be maintained.

(2) Observe the other boat against its background, which may be land, stars or even a buoy. If the boat is stationary against the background and remains so, a collision will occur. If, on the other hand, the vessel is moving ahead of, or falling back on the background, all is well.

The full text of the IRPCS is obtainable from many sources: the prudent skipper and all keen sailors should obtain their own copy. The RYA produce a pocket-sized booklet at very low cost, and it is recommended by this author.

INDEX